Journeyman Electrician's Review

THIRD EDITION

Richard E. Loyd

Delmar
Thomson Learning™

Africa • Australia • Canada • Denmark • Japan • Mexico
New Zealand • Phillipines • Puerto Rico • Singapore
Spain • United Kingdom • United States

NOTICE TO THE READER

Delmar Staff

Business Unit Director: Alar Elken
Acquisitions Editor: Mark Huth
Editorial Assistant: Dawn Daugherty
Executive Marketing Manager: Maura Theriault
Channel Manager: Mona Caron

Executive Production Manager: Mary Ellen Black
Project Editor: Barbara Diaz
Production Coordinator: Toni Hansen
Art/Design Coordinator: Rachel Baker
Cover Design: Brucie Rosch

COPYRIGHT © 2000
Delmar is a division of Thomson Learning. The Thomson Learning logo is a registered trademark used herein under license.

Printed in the United States of America
1 2 3 4 5 6 7 8 9 10 XXX 05 04 03 02 01 00 99

For more information, contact Delmar, 3 Columbia Circle, PO Box 15015, Albany, NY 12212-5015; or find us on the World Wide Web at http://www.delmar.com

Asia
Thomson Learning
60 Albert Street
#15–01 Albert Complex
Singapore 189969
Tel: 65 336 6411
Fax: 65 336 7411

Japan
Thomson Learning
Palaceside Building 5F
1-1-1 Hitotsubashi, Chiyoda-ku
Tokyo 100 0003 Japan
Tel: 813 5218 6544
Fax: 813 5218 6551

Australia/New Zealand
Nelson/Thomson Learning
102 Dodds Street
South Melbourne, Victoria 3205
Australia
Tel: 61 39 685 4111
Fax: 61 39 685 4199

UK/Europe/Middle East
Thomson Learning
Berkshire House
168–173 High Holborn

London
WC1V7AA United Kingdom
Tel: 44 171 497 1422
Fax: 44 171 497 1426

Thomas Nelson & Sons Ltd.
Nelson House
Mayfield Road
Walton-on-Thames
KT 12 5PL United Kingdom
Tel: 44 1932 2522111
Fax: 44 1932 246574

Latin America
Thomson Learning
Seneca, 53
Colonia Polanco
11560 Mexico D.F. Mexico
Tel: 525-281-2906
Fax: 525-281-2656

South Africa
Thomson Learning
Zonnebloem Building
Constantia Square
526 Sixteenth Road
PO Box 2459
Halfway House, 1685

South Africa
Tel: 27 11 805 4819
Fax: 27 11 805 3648

Canada
Nelson/Thomson Learning
1120 Birchmount Road
Scarborough, Ontario
Canada M1K 5G4
Tel: 416-752-9100
Fax: 416-752-8102

Spain
Thomson Learning
Calle Magallanes, 25
28015-Madrid
España
Tel: 34 91 446 33 50
Fax: 34 91 445 62 18

International Headquarters
Thomson Learning
International Division
290 Harbor Drive, 2nd Floor
Stamford, CT 06902-7477
Tel: 203-969-8700
Fax: 203-969-8751

Library of Congress Cataloging-in-Publication Data

Loyd, Richard E.
    Journeyman electrician's review / Richard E. Loyd.—3rd ed.
        p.   cm.
    Includes index.
    ISBN 0-7668-1277-4 (alk. paper)
    1. Electric engineering—Examinations, questions, etc.   2. Electric
engineering Miscellanea.   3. Electricians—Licenses—United States.
I. Title.
TK169.L69   1999
621.319′24—dc21   99-14460
                                CIP

Dedication

I wish to dedicate this book to my wife and business partner, Nancy. She is my inspiration and number one fan. Her assistance is invaluable. I would also like to thank all my wonderful circuit rider friends. Their expertise in the *National Electrical Code*® and the electrical industry continue to provide me with the latest information on our great industry. Finally, to Earl Featherston, my apprenticeship instructor in Pocatello, Idaho, IBEW local union 449, almost forty years ago. Earl put up with us and taught us from the first class to the last class that we must be able to understand and apply the rules of the *National Electrical Code*®.

Online Services

Delmar Online
To access a wide variety of Delmar products
and services on the World Wide Web, point your
browser to:
http://www.delmar.com
or email: info@delmar.com

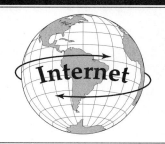

Delmar, Thomson Learning Is Your Electrical Book Source!

Whether you're a beginning student or a master electrician, Delmar, Thomson Learning has the right book for you. Our complete selection of proven best-sellers and all-new titles is designed to bring you the most up-to-date, technically accurate information available.

NATIONAL ELECTRICAL CODE®

National Electrical Code® 1999/ NFPA
Revised every three years, the *National Electric Code®* is the basis of all U.S. electrical codes.
Order # 0-8776-5432-8
Loose-leaf version in binder
Order # 0-8776-5433-6

National Electrical Code® Handbook 1999/ NFPA
This essential resource pulls together all the extra facts, figures, and explanations you need to interpret the 1999 *NEC®*.
It includes the entire text of the Code, plus expert commentary, real-world examples, diagrams, and illustrations that clarify requirements.
Order # 0-8776-5437-9

Illustrated Changes in the 1999 National Electrical Code® / O'Riley
This book provides an abundantly-illustrated and easy-to-understand analysis of the changes made to the 1999 *NEC®*.
Order # 0-7668-0763-0

Understanding the National Electrical Code®, 3E / Holt
This book gives users at every level the ability to understand what the *NEC®* requires, and simplifies this sometimes intimidating and confusing Code.
Order # 0-7668-0350-3

Illustrated Guide to the National Electrical Code® / Miller
Highly detailed illustrations offer insight into Code requirements, and are further enhanced through clearly written, concise blocks of text that can be read very quickly and understood with ease. Organized by classes of occupancy.
Order # 0-7668-0529-8

Interpreting the National Electrical Code®, 5E / Surbrook
This updated resource provides a process for understanding and applying the *National Electrical Code®* to electrical contracting, plan development, and review.
Order # 0-7668-0187-X

Electrical Grounding, 5E / O'Riley
Electrical Grounding is a highly illustrated, systematic approach for understanding grounding principles and their application to the 1999 *NEC®*.
Order # 0-7668-0486-0

ELECTRICAL WIRING

Electrical Raceways and Other Wiring Methods, 3E / Loyd
The most authoritive resource on metallic and nonmetallic raceways, provides users with a concise, easy-to-understand guide to the specific design criteria and wiring methods and materials required by the 1999 *NEC®*.
Order # 0-7668-0266-3

Electrical Wiring Residential, 13E / Mullin
Now in full color! Users can learn all aspects of residential wiring and how to apply them to the wiring of a typical house from this, the most widely used residential wiring book in the country.
Softcover Order # 0-8273-8607-9
Hardcover Order # 0-8273-8610-9

House Wiring with the NEC® / Mullin
The focus of this new book is the applications of the *NEC®* to house wiring.
Order # 0-8273-8350-9

Electrical Wiring Commercial, 10E / Mullin and Smith
Users can learn commercial wiring in accordance with the *NEC®* from this comprehensive guide to applying the newly revised 1999 *NEC®*.
Order # 0-7668-0179-9

Electrical Wiring Industrial, 10E / Smith and Herman
This practical resource has users work their way through an entire industrial building—wiring the branch-circuits, feeders, service-entrances, and many of the electrical appliances and sub-systems found in commercial buildings.
Order # 0-7668-0193-4

Cables and Wiring, 2E / AVO
This concise, easy-to-use book is your single-source guide to electrical cables—it's a "must-have" reference for journeyman electricians, contractors, inspectors, and designers.
Order # 0-7668-0270-1

ELECTRICAL MACHINES AND CONTROLS

Industrial Motor Control, 4E / Herman and Alerich
This newly revised and expanded book, now in full color, provides easy-to-follow instructions and essential information for controlling industrial motors. Also available are a new lab manual and an interactive CD-ROM.
Order # 0-8273-8640-0

Electric Motor Control, 6E / Alerich and Herman
Fully updated in this new sixth edition, this book has been a long-standing leader in the area of electric motor controls.
Order # 0-8273-8456-4

Introduction to Programmable Logic Controllers / Dunning
This book offers an introduction to Programmable Logic Controllers.
Order # 0-8273-7866-1

Technician's Guide to Programmable Controllers, 3E / Cox
Uses a plain, easy-to-understand approach and covers the basics of programmable controllers.
Order # 0-8273-6238-2

Programmable Controller Circuits / Bertrand
This book is a project manual designed to provide practical laboratory experience for one studying industrial controls.
Order # 0-8273-7066-0

Electronic Variable Speed Drives / Brumbach
Aimed squarely at maintenance and troubleshooting, *Electronic Variable Speed Drives* is the only book devoted exclusively to this topic.
Order # 0-8273-6937-9

Electrical Controls for Machines, 5E / Rexford
State-of-the-art process and machine control devices, circuits, and systems for all types of industries are explained in detail in this comprehensive resource.
Order # 0-8273-7644-8   *(Continued on next page)*

Electrical Transformers and Rotating Machines / Herman
This new book is an excellent resource for electrical students and professionals in the electrical trade.
Order # 0-7668-0579-4

Delmar's Standard Guide to Transformers / Herman
Delmar's Standard Guide to Transformers was developed from the best-seller *Standard Textbook of Electricity* with expanded transformer coverage not found in any other book.
Order # 0-8273-7209-4

DATA AND VOICE COMMUNICATION CABLING AND FIBER OPTICS

Complete Guide to Fiber Optic Cable System Installation / Pearson
This book offers comprehensive, unbiased, state-of-the-art information and procedures for installing fiber optic cable systems.
Order # 0-8273-7318-X

Fiber Optics Technician's Manual / Hayes
Here's an indispensable tool for all technicians and electricians who need to learn about optimal fiber optic design and installation as well as the latest troubleshooting tips and techniques.
Order # 0-8273-7426-7

A Guide for Telecommunications Cable Splicing / Highhouse
A "how-to" guide for splicing all types of telecommunications cables.
Order # 0-8273-8066-6

Premises Cabling / Sterling
This reference is ideal for electricians, electrical contractors, and inspectors needing specific information on the principles of structured wiring systems.
Order # 0-8273-7244-2

ELECTRICAL THEORY

Delmar's Standard Textbook of Electricity, 2E / Herman
This exciting full-color book is the most comprehensive book on DC/AC circuits and machines for those learning the electrical trades.
Order # 0-8273-8550-1

Industrial Electricity, 6E / Nadon, Gelmine, and Brumbach
This revised, illustrated book offers broad coverage of the basics of electrical theory and industrial applications. It is perfect for those who wish to be industrial maintenance technicians.
Order # 0-7668-0101-2

EXAM PREPARATION

Electrician's Exam Preparation / Holt
This comprehensive exam prep guide includes all of the topics on both the master and journeyman electrician competency exams.
Order # 0-7668-0376-7

REFERENCE

ELECTRICAL REFERENCE SERIES
This series of technical reference books is written by experts and designed to provide the electrician, electrical contractor, industrial maintenance technician, and other electrical workers with a source of reference information about virtually all of the electrical topics that they encounter.

Electrician's Technical Reference—Motor Controls / Carpenter
Electrician's Technical Reference—Motor Controls is a source of comprehensive information on understanding the controls that start, stop, and regulate the speed of motors.
Order # 0-8273-8514-5

Electrician's Technical Reference—Motors / Carpenter
Electrician's Technical Reference—Motors builds an understanding of the operation, theory, and applications of motors.
Order # 0-8273-8513-7

Electrician's Technical Reference—Theory and Calculations / Herman
Electrician's Technical Reference—Theory and Calculations provides detailed examples of problem-solving for different kinds of DC and AC circuits.
Order # 0-8273-7885-8

Electrician's Technical Reference—Transformers / Herman
Electrician's Technical Reference—Transformers focuses on the theoretical and practical aspects of single-phase and 3-phase transformers and transformer connections.
Order # 0-8273-8496-3

Electrician's Technical Reference—Hazardous Locations / Loyd
Electrician's Technical Reference—Hazardous Locations covers electrical wiring methods and basic electrical design considerations for hazardous locations.
Order # 0-8273-8380-0

Electrician's Technical Reference—Wiring Methods / Loyd
Electrician's Technical Reference—Wiring Methods covers electrical wiring methods and basic electrical design considerations for all locations, and shows how to provide efficient, safe, and economical applications of various types of available wiring methods.
Order # 0-8273-8379-7

Electrician's Technical Reference—Industrial Electronics / Herman
Electrician's Technical Reference—Industrial Electronics covers components most used in heavy industry, such as silicon control rectifiers, triacs, and more. It also includes examples of common rectifiers and phase-shifting circuits.
Order # 0-7668-0347-3

RELATED TITLES

Common Sense Conduit Bending and Cable Tray Techniques / Simpson
Now geared especially for students, this manual remains the only complete treatment of the topic in the electrical field.
Order # 0-8273-7110-1

Practical Problems in Mathematics for Electricians, 5E / Herman
This book details the mathematics principles needed by electricians.
Order # 0-8273-6708-2

Electrical Estimating / Holt
This book provides a comprehensive look at how to estimate electrical wiring for residential and commercial buildings with extensive discussion of manual versus computer-assisted estimating.
Order # 0-8273-8100-X

Electrical Studies for Trades / Herman
Based on *Delmar's Standard Textbook of Electricity*, this new book provides non-electrical trades students with the basic information they need to understand electrical systems.
Order # 0-8273-7845-9

Table of Contents

Foreword

ELECTRIC PER SE

Electricity! In ancient times it was believed to be an act of the gods. Not even Greco-Roman civilizations could understand this thing we know as electricity. It was not until about 1600 A.D. that any scientific theory was recorded; after seventeen years of research, William Gilbert wrote a book on the subject titled *De Magnete*. It was almost another 150 years before major gains were made in understanding electricity so people might some day be able to harness it. At that time Benjamin Franklin, often referred to as the grandfather of electricity or electrical science, and his close friend Joseph Priestley, a great historian, began to gather the works of the many worldwide scientists who had been working independently. (One has to wonder where this wondrous industry would be today if the great minds of yesteryear had had the benefit of the great communication networks we have today.) Benjamin Franklin traveled to Europe to gather this information, and with that trip the industry began to surge forward for the first time. Franklin's famous kite experiment allegedly took place in about 1752. Coupling the results of that with other

scientific data gathered, Franklin decided he could sell installations of lightning protection to every building owner in the city of Philadelphia. He was very successful in doing just that! (We credit Franklin for the terms *conductor* and *nonconductor*, replacing *electric per se* and *non-electric*.)

With the accumulation of the knowledge gained from these many scientists and inventors of the past, many other great minds emerged and lent their names to even more discoveries. Their names are as familiar as the terms related to electricity. Alessandro Volta, James Watt, André Marie Ampère, George Ohm, and Heinrich Hertz are names of just a few of the great minds that enabled the electrical industry to come together.

The next major breakthrough came only a little more than 100 years ago. Thomas Edison, the holder of hundreds of patents promoting direct current (DC) voltage, and Nikola Tesla, inventor of the 3-phase motor and the promoter of alternating current (AC) voltage, began an inimical (bitter) competition in the late 1800s. Edison was methodically making in-roads with DC current all along the Eastern Seaboard, and

Electricity is here! *(Courtesy of Underwriters Laboratories, Inc.)*

1893 Great Columbian Exposition at the Chicago World's Fair. The Palace of Electricity astonished crowds; it also astonished electricians by repeatedly setting fire to itself. *(Courtesy of Underwriters Laboratories, Inc.)*

Tesla had exhausted his finances for his experimental project with AC current in Colorado. Then, the city of Chicago requested bids to light the great Columbian Exposition of 1893. Lighting the World's Fair in Chicago, celebrating 300 years since Columbus discovered America, required building generators and nearly a quarter of a million lights. Edison teamed up with General Electric and submitted a bid based on the Edison light bulb powered by DC current. George Westinghouse, already successful in the electrified railroad, the elevator, and about four hundred other inventions, joined with Nikola Tesla and submitted the lowest bid—about one-third of the Edison-General Electric bid. Westinghouse and Tesla were awarded the contract. They based their bid on the Westinghouse Stopper Light powered by alternating current. The installation was made by Guarantee Electric of Saint Louis, Missouri, which is now one of the oldest and largest electrical contractors in the world. It is said that thousands came out to see this great lighting display. Almost as spectacular as the lighting were the arcing, sparking, and fires started by the display. Accidents occurred daily! It is said that Edison embellished the dangers of AC current by electrocuting dogs and promoted DC current as being much safer. However, the public knew better, since plenty of fires and accidents also occurred with DC current.

About this time, three very important events took place almost simultaneously. The Chicago Board of Fire Underwriters hired William Henry Merrill, an electrician from Boston, as an electrical inspector for the exposition. Merrill saw the need for safety inspections and started Underwriters Laboratories in 1894. In 1895, the first nationally recommended Code was published by the National Board of Fire Underwriters (now the American Insurance Association). The 1895 *National Electrical Code®* was drafted through the combined efforts of the architectural, electrical, insurance, and other allied interests based on a document that resulted from actions taken in 1892 by Underwriters National Electrical Association, which met and consolidated many various Codes that were already in existence. The first *National Electrical Code®* was presented to the National Conference on Electrical Rules, which was composed of delegates from various national associations who voted to unanimously recommend it to their respective associations for adoption or approval. A group from Buffalo, New York, visited the World's Fair and, based on that visit, hired George Westinghouse and Nikola Tesla to build AC hydro-generators to be powered by Niagara Falls. Alternating current had won out over direct current as this country's major power source. The rest is history!

Preface

Although the primary purpose of *Journeyman Electrician's Review* is to provide electrical students with a concise, easily understandable study guide to the 1999 edition of the *National Electrical Code*® and the application of electrical calculations, it is written recognizing that the electrical field has many facets and the user may have diverse interests and varying levels of experience. However, the interests of the relatively experienced electrician preparing to enter the electrical industry as an entry-level journeyman electrician, for the electrician in training, and especially for the advanced electrical student apprentice, have been considered carefully in the text's preparation. *Journeyman Electrician's Review* provides a ready source of the basic information on the *NEC*®. This text may be used for reference or as study material for students and electricians who are preparing for an examination for licensing. It is especially designed as an aid to those studying for the nationally recognized examinations, such as the National Assessment Institute (NAI) and Block and Associates Inc. It will also help anyone preparing for any electrical examination and will provide a quick, easily understood study guide for those needing to update themselves on the *National Electrical Code*® and basic electrical mathematical formulas and calculations. This book covers basic calculations; however, for the study of complex electrical calculations *Understanding Electrical Calculations* by Mike Holt, Delmar, Thomson Learning, is recommended. The text is brief and concise for easy application for classroom or home study.

Each lesson is designed purposely to require the student to apply the entire *NEC*® text, and not specific chapters or articles. It has been found that when studying for a timed, open-book examination, the student must gain proficiency in the Table of Contents, the Index, and the ability to move quickly from cover to cover to find the correct answer to each question in a timely fashion. Check with the testing organization to see if you are permitted to add index tabs or highlight important sections. If permitted, this can speed up your search for the information.

STUDY RECOMMENDATIONS

- Read Chapter One carefully.
- Complete the sample examination. The most important point is that you treat the test the same way you would the "real test." Allow yourself three hours, and grade yourself honestly. This examination will point out to you the areas in which you are weak, i.e., speed, grounding, or services.
- Study Chapters Two through Thirteen and complete the question review sections following each chapter.
- Complete the fifty-question practice examination in Chapter Fourteen. Allow yourself two hours for the examination.
- Evaluate the areas in which you are still weak.
- Complete the practice exams in Chapter Fourteen until you feel you have overcome your weak areas.
- Complete the final examination in Chapter Fourteen. Allow yourself three hours. If you have studied, you should be prepared. Good luck!

ACKNOWLEDGMENTS

The author would like to give special thanks to the following companies and their representatives for their help and contributions:

The National Assessment Institute
Crouse-Hinds
Alflex
Ray Mullin, formerly with Bussmann Manufacturing, Division of Cooper Industries
Patricia H. Horton, Allied Tube and Conduit Company
Charles Forsberg, Carlon, A Lampson and Sessions Company
Bill Slater, Raco Incorporated, Subsidiary of Hubbell Inc.
Jim Pauley, Square D Company
Mike Holt, Mike Holt Enterprises
Roger Sanstedt, B-Line Systems Inc.

The author and Delmar, Thomson Learning would like to acknowledge and thank Edward Clancy of Sarasota County Technical Institute for reviewing this manuscript.

Applicable tables and section references are reprinted with permission from NFPA 70-1999, *National Electrical Code, NEC®* copyright © 1998, National Fire Protection Association, Quincy, MA 02269. This reprinted material is not the complete and official position of the NFPA on the referenced subject, which is represented by the standard in its entirety.

About the Author

Richard E. Loyd is a nationally known author and consultant specializing in the *National Electrical Code®* and the model building codes. He is president of his own firm, R & N Associates, located in Perryville, Arkansas. He and his wife Nancy travel throughout the country presenting seminars and speaking at industry-related conventions. He also serves as a code expert at 35 to 40 meetings per year at the International Association of Electrical Inspectors (IAEI) meetings throughout the United States. Mr. Loyd represents the Steel Tube Institute of North America (STI) as an *NEC®* consultant. He represents the American Iron and Steel Institute (AISI) on ANSI/NFPA 70 in the *National Electrical Code®* as a member of Code Committee Eight, the panel responsible for raceways. He is currently immediate past Chairman of the National Board of Electrical Examiners (NBEE). Mr. Loyd is actively involved in forensic inspections and investigations on a consulting basis and serves as an expert witness in matters related to codes and safety. He is vice-chairman and an active member of the Electrical Section of the National Fire Protection Association (NFPA) and an active member of the Arkansas Chapter of Electrical Inspectors Association Executive Board (IAEI). He is a contributing editor for Intertec Publications *(EC&M Magazine)*.

He is an active member in the Institute of Electrical & Electronics Engineers Inc. (IEEE), where he serves on the Power Systems Grounding Committee (Green Book), the International Conference of Building Officials (ICBO), the Southern Building Code Congress International (SBCCI), and the Building Officials & Code Administrators International (BOCA). Mr. Loyd is currently licensed as a master contractor/electrician in Arkansas (License #1725) and Idaho (License #2077), and is an NBEE-certified master electrician.

Mr. Loyd served as the Chief Electrical Inspector and Administrator for the State of Idaho and for the State of Arkansas. He has served as Chairman of NFPA 79 "Electrical Standard for Industrial Machinery," as a member of Underwriters Laboratories Advisory Electrical Council, as Chairman for Educational Testing Service (ETS), a multistate electrical licensing advisory board, and as a master electrician and electrical contractor. He has been accredited to teach licensing certification courses in Florida, Idaho, Oregon, and Wyoming, and has taught basic electricity and *National Electrical Code®* classes for Boise State University (BSU) in Idaho.

Chapter One

EXAMINATIONS AND NATIONAL TESTING ORGANIZATIONS

Examinations are an old phenomenon. They have been given to many disciplines to evaluate one's competency in their field of specialty. Until recently they were prepared by others within their chosen field, based on the knowledge of each person preparing the examination. Today many examinations are professionally developed by state agencies specializing in examination preparation and national testing organizations.

In recent years many local jurisdictions—city, county, and state—have dropped their electrical examinations in favor of the national testing agencies that prepare and administer electrical examinations for those local entities. However, there are still many local jurisdictions developing and administering electrical examinations much in the same manner as has been done since the beginning of electrical examinations. These examinations are often prepared and administered by local electrical inspectors and are often developed with question material related to areas in which the inspector often finds violations or to areas in the *National Electrical Code®* where the inspector feels are most difficult to understand. Rarely do local examinations fully evaluate an entry-level craftsman to his or her competency in this new career field. National testing organizations, such as the National Assessment Institute–Block & Associates (NAI–Block), have carefully taken the advantages of localized testing methodology and incorporated the latest state-of-the-art technology to develop a test that is fair to the taker as well as a good evaluation of the taker's knowledge in this specific career field. These testing organizations use those experienced in the trades in pools and in sit-down committees to develop a task analysis to determine what the examination will evaluate and to then develop questions from throughout those tasks. These organizations strive to develop question criteria that is clear and not confusing, questions that have only one answer, and questions in which an entry-level candidate into the journeyman electrical field or into the master contractor field should know as minimum standards to enter those fields. These examinations are generally multiple choice. The study arrangement in this book is designed to prepare the user to successfully pass these national examinations. However, it also should provide those preparing for localized tests on these same subjects.

DEVELOPING AN ITEM (QUESTION) BANK

The development of examinations that effectively measure the minimum competency of an entry-level candidate is a process that uses many concepts. National testing firms, such as the National Assessment Institute–Block & Associates and formerly ETS (the Electrical Testing Service for multistate licensing), have developed examinations that have become a standard in the industry today. The first step in developing an examination is a task or job analysis. The task analysis must evaluate the ability, knowledge, and skill to perform the tasks related to the job in a way that will not endanger public health, safety, and welfare. They must be relevant to the actual electrical practice in the field. The first step is then to establish the criteria, the subject areas of the trade that are most important and need to be tested. An acceptable method for determining these various job-related tasks would be a survey of licensed tradespeople performing these tasks in the field on a day-to-day basis, or an assembly of competent experts in these fields meeting in a forum to list the content outline and determining the degree of importance of each task to be evaluated by the examination, a blueprint that shows which percent of the exam for each subject task. This blueprint ties the test to the job performance, and the test outline shows the importance of each subject. When done properly, the task analysis will substantiate the validity of the exam.

The next task is to hold workshops or group meetings to develop an item (question) bank for a group of questions. This task of developing the item bank is normally performed by a group of interested experts working in workshops to develop the item bank based on the *National Electrical Code®* and other electrical related texts or reference books. As the item bank is developed, it is important that each item be clear, precise, and have only one correct answer. On completion of the item bank, there should be a pool of 500–800 questions. The questions are entered into the computer, each file based on the task analysis for each job task and weighted according as to the difficulty level of each question. The electrical board developed from contract states subscribing to these national tests then determines the passing scores, the item selection, and test form assembly. When this has been completed, the examination is ready for administration. Following the test administration to the candidates, a post-test examination analysis is then conducted to determine the effectiveness of the examination. In the post-test analysis, which is conducted following every exam, the effectiveness of each item is examined. Inadequate items are returned to the workshop to be rewritten or corrected or are eliminated from future exams. Each item is evaluated from developed criteria from previous exams, and the difficulty level is verified. Postexamination analysis provides critical and important information for future workshops and the validity of the exams. This analysis provides information that cannot be gained in any other manner. Information continues to be developed throughout the life of the item bank. This information is fed into the computer and the continually proves and improves the validity of future examinations. The candidates themselves provide the most important information in the continuing development and improvement of the examination.

PREPARING FOR AN EXAMINATION

The first step in preparing for an examination is to obtain the examination information from the Authority Having Jurisdiction to which you are about to make application for an examination. National testing firms generally furnish each authority having jurisdiction a bulletin containing generic material that applies nationwide, with specific unique material related only to the jurisdiction in which you are about to take an examination. This bulletin contains information that must be read carefully about the eligibility and procedure for registering for the test. This includes your verification of experience and education requirements and usually requires an application fee. It also contains the deadlines that are important for submitting the application and fees, and the examination dates related. For instance, you may be required to have your application and fees submitted up to 60 days before the examination date. The fees vary from state to state. They generally require

the application fee be made to the jurisdiction and the exam administration fee to be made directly to the national testing firm. However, this procedure varies and must be read carefully. Once your application has been approved, your name will be placed on the roster for the next administration of the exam in your area. The schedule of examination dates and locations is available through the local jurisdiction. You will be mailed a registration form that will serve as an admission letter. Do not lose this because it will be required when you arrive at the test center or test location to take the exam. If you fail to bring this letter, you will not be permitted to take the exam. Also, it is very important that you read this bulletin to determine what materials are required that you bring and what materials are allowed, usually the *National Electrical Code®* or handbook in the edition in which the test is developed. Other electrical reference books may be permitted. You are usually required to bring your own pencils and a silent, handheld, nonprogrammable, battery-powered calculator. Some jurisdictions also require a hands-on practical test. If it is required in the area in which you are making application, the outline of that work should also appear in the bulletin. Read it carefully because the practical test can mean the difference between passing and failing the examination. One such jurisdiction required that the candidate bring conduit, a conduit bender, hacksaw, reaming tool, tape measure, and continuity or ohm-meter. Also contained in the candidate bulletin will be the requirements for candidates with physical disabilities. These disabilities must be submitted with documentation at the time the application is made. Special considerations are generally permitted to accommodate religious or disability needs in accordance with the American Disabilities Act (ADA).

> **Caution:** Verify which *NEC®* edition is being used (1996 or 1999). Check allowable reference books before studying for the examination. Some agencies permit only the softcover *NEC®* while others permit the *NFPA® Handbook* and other reference books. Verify what pencil type and calculator types are allowed. Do not wait until test day. Be prepared! A photo ID may be required.

Finally, in the candidate bulletin is a content outline, the number of questions in each content area, the study reference material that the test has been developed from, such as the *National Electrical Code®* book; *Alternating Current Fundamentals*, Duff & Herman, Delmar, Thomson Learning; *Direct Current Fundamentals*, Loper &

Tedsen, Delmar, Thomson Learning; the *American Electrician's Handbook*; and a sample of questions. It is suggested that you cover the answers and take this sample test in much the same manner as you would take the real examination. It gives you an idea of what the test will be like and what the format of the test is if you were not previously familiar with multiple-choice examinations. This is an important part of preparing for the electrical examination. Take the candidate bulletin seriously.

The next step in preparing for the examination is to study this book. The review questions at the end of each chapter are taken from throughout the *National Electrical Code* and not necessarily only from the content of each chapter. The *National Electrical Code* is a reference document. As a reference document, it is necessary that you obtain a good proficiency in using the Table of Contents and the Index to be able to quickly locate the article and section of the Code. For each review question that you answer, you should not only research and find the correct answer, but you should also reference the *NEC* section in which the answer was found. As you go through all the review material, take the sample tests throughout the document. This will enable you to pass the test the first time you take it. Anything less may prove disappointing. You should be aware that if you fail the examination, many jurisdictions require a waiting period of six months to one year before you can retake the exam.

The next section of this book contains a sample examination similar to the examination that you will take in the jurisdiction that you are studying for and similar to the examination given by many national testing organizations. This test should be taken in a quiet room. You should time yourself for four hours. It should be taken all at one time. To do otherwise will not give you a true evaluation of your present knowledge and of the areas in which you are weakest where you need to concentrate your efforts and study time to gain the competency needed to pass your examination.

You may have purchased this book for various reasons. Some individuals study merely to improve their understanding in their chosen field, but most do so because they are required to do so to pass their performance evaluation on their job or to pass an evaluation in a state or local examination. The specific reason is unimportant. The important thing is that you are about to undertake the study of electricity that, if done properly, will improve your competency in this field. The better you prepare, the better you will do on any evaluation or examination. Preparedness has more than one effect. First of all, your IQ can fluctuate 30 to 40 points

between any given date. The best way to improve an individual's performance is to reduce the anxiety level. This book and other Delmar, Thomson Learning books related to this subject are designed to improve an individual's understanding on how to use the *National Electrical Code* as a reference document, how to use the basic electrical formulas as applicable, and to duplicate as nearly as possible the test format used by many national testing organizations and local jurisdictions. Whether you are preparing for an examination or just to improve your skill, good study habits will give you the maximum efficient use of your time. To improve your study habits, consistent study time is important. You should allow yourself the same amount of time each day or each week to study this course. Consistent study time improves your retention.

Test Day

The time has finally arrived for you to take your examination. Be sure you allow plenty of time to arrive on time because some testing agencies will not admit you if you are late, but do not arrive too early. Discussing the test with other examinees may raise your anxiety level. Nervousness is contagious! Be sure you have your receipt for your application and your entrance ticket; be sure you have all materials required by the jurisdiction or testing agency, including study books and a hand-held calculator.

When a test is multiple choice, you generally have four alternative answers. The odds are four to one that you can get the right answer. If you can eliminate two of the alternatives, your chances are now increased to 50/50. This is extremely beneficial when referencing the Code, as you can quickly eliminate half of the possibilities leaving you only two to research.

Read the directions carefully. Many mistakes are made merely because of misunderstanding. If you have any questions, discuss them with the proctor before the test time starts. Any discussions after the test starts will take away from the time you have to answer the questions. Allow yourself ample time for each question. Answer the question first before you check any alternatives. This way you can evaluate your answer against any alternatives that you may have found in the *NEC*. Do not spend too much time on any one question. You should never spend more than $3\frac{1}{2}$ minutes on a question. If you find the question extremely difficult and you are unsure of the answer or cannot find it in the *NEC*, skip it and only go back to it if you have time at the end of the test. Your first choice is usually best. Read the questions very carefully. Make sure you know key words that might change the meaning of the question.

Note any negatives, such as "Which of the following are *not* . . . ," ". . . shall *not* be permitted . . ." It may be beneficial to underline key words as you read the question. This has a tendency to channel your thoughts in the right direction. Dress properly. The test facility should be approximately 70°F. Too many clothes or uncomfortable clothes have a tendency to make you drowsy and makes concentration difficult. After you have answered all the questions, recheck your work. Make sure you have not made clerical mistakes. Make sure your answer is recorded correctly on the answer sheet. Check the numbers against each other periodically. Go over the extremely difficult questions. Questions on which you know you have guessed you may want to mark in some way so if there is time at the end of the test, you can do additional research. Good luck!

SAMPLE EXAMINATION

Examination Instructions:

For a positive evaluation of your knowledge and preparation awareness, you must:

1. Locate yourself in a quiet atmosphere (room by yourself).

2. Have with you at least two sharp No. 2 pencils, the 1999 *NEC®*, and a hand-held calculator.

3. Time yourself (three hours), with no interruptions.

4. After the test is complete, grade yourself honestly and concentrate your studies on the sections of the *NEC®* in which you missed the questions.

   Caution: Do not just look up the correct answers, as the questions in this examination are only an exercise and not actual test questions. Therefore, it is important that you be able to quickly find answers from throughout the *NEC®*.

Each lesson is designed purposely to require the student to apply the entire *NEC®* text, not specific chapters or articles. It has been found that when studying for a timed, open-book examination the student must gain proficiency in the Table of Contents, the Index, and the ability to move quickly from cover to cover to find the correct answers to each question in a timely fashion.

Directions: Each question is followed by four suggested answers A, B, C, and D. In each case select the **one** that best answers the question.

1. Liquidtight flexible metal conduit can be used in which of the following locations?
   A. In areas that are both exposed and concealed
   B. In areas that are subject to physical damage
   C. In connection areas for gasoline dispensing pumps
   D. In areas where ambient temperature is to be greater than 200°C

   Answer: _____   Reference: _____

2. The maximum number of quarter bends allowed in one run of nonmetallic rigid conduit is
   A. 2
   B. 4
   C. 6
   D. 8

   Answer: _____   Reference: _____

3. The maximum size of flexible metallic tubing that can be used in any construction or installation is
   A. ⅜ inch
   B. ½ inch
   C. ¾ inch
   D. 1 inch

   Answer: _____   Reference: _____

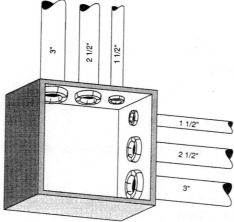

4. The size of the pull box in the diagram above should not be less than
    A. 12 inches × 14 inches
    B. 16 inches × 18 inches
    C. 22 inches × 22 inches
    D. 24 inches × 24 inches

    Answer: _____   Reference: _____

5. A metal raceway system installed in the washing area of a car wash must be spaced at least how far from the walls?
    A. ⅛ inch
    B. ¼ inch
    C. ¾ inch
    D. 1 inch

    Answer: _____   Reference: _____

6. Unless otherwise indicated, busways should be supported at intervals not to exceed how many feet?
    A. 3 feet
    B. 5 feet
    C. 7 feet
    D. 10 feet

    Answer: _____   Reference: _____

7. Which of the following percentages of conduit fill should be used when four type THW conductors are being installed?
    A. 55 percent
    B. 53 percent
    C. 40 percent
    D. 31 percent

    Answer: _____   Reference: _____

8. Flexible appliance cords of 10 amp capacity are considered adequately protected if the circuit overcurrent device is set at a maximum value of how many amperes?
    A. 15
    B. 20
    C. 30
    D. 40

    Answer:_____   Reference:_____

9. All of the following sizes of solid aluminum conductors shall be made of an aluminum alloy except
    A. No. 8 AWG
    B. No. 10 AWG
    C. No. 12 AWG
    D. No. 14 AWG

    Answer:_____   Reference:_____

10. An insulated bushing is required to be used on a raceway entering a cabinet if the ungrounded conductors entering the raceway are at least
    A. No. 10 AWG
    B. No. 8 AWG
    C. No. 6 AWG
    D. No. 4 AWG

    Answer:_____   Reference:_____

11. The allowable distance between supports of nonmetallic-sheathed cable installed in an on-site constructed one-family dwelling is a maximum of how many feet?
    A. 2½ feet
    B. 3 feet
    C. 4 feet
    D. 4½ feet

    Answer:_____   Reference:_____

12. It is permissible to use No. 4 THW CU AWG for service in dwelling units to a maximum load of
    A. 100 amps
    B. 95 amps
    C. 70 amps
    D. 65 amps

    Answer:_____   Reference:_____

13. Where conductors of less than 600 volts emerge from the ground, they must be protected by enclosures or raceways that extend from below grade to a point how many feet above the finish grade?
    A. 5 feet
    B. 6 feet
    C. 7 feet
    D. 8 feet

    Answer:_____   Reference:_____

14. The service rating for a residence that uses No. 2 THW copper service-entrance conductors is
    A. 95 amps
    B. 110 amps
    C. 115 amps
    D. 125 amps

    Answer: _____   Reference: _____

15. Surge arrester grounding conductors that run in metal enclosures should be
    A. bare
    B. insulated
    C. bonded on one end of the enclosure only
    D. bonded at both ends of the enclosure

    Answer: _____   Reference: _____

16. The number of overcurrent devices in a single cabinet of a lighting and appliance panelboard should not exceed
    A. 30
    B. 36
    C. 42
    D. 48

    Answer: _____   Reference: _____

17. When it is impractical to locate a service head directly above the point of attachment, the maximum allowable placement distance from the point of attachment is how many feet?
    A. 1 foot
    B. 2 feet
    C. 3 feet
    D. 4 feet

    Answer: _____   Reference: _____

18. An underground service to a small controlled water heater is to be installed. The single branch-circuit used for this service requires that copper conductors be at least
    A. No. 12 AWG
    B. No. 10 AWG
    C. No. 8 AWG
    D. No. 6 AWG

    Answer: _____   Reference: _____

19. Service conductors can have less than 3 feet of clearance from which of the following?
    A. Doors
    B. Tops of windows
    C. Fire escapes
    D. Porches

    Answer: _____   Reference: _____

20. Service conductors that pass over rooftops should have a vertical clearance of not less than how many feet?
    A. 3 feet
    B. 7 feet
    C. 8 feet
    D. 10 feet

    Answer: _____   Reference: _____

21. The interruptive rating shall be marked on all circuit breakers rated more than
    A. 1,000 amps
    B. 2,000 amps
    C. 2,500 amps
    D. 5,000 amps

    Answer: _____   Reference: _____

22. Illumination is mandatory for service equipment or panelboards in a dwelling unit if the service to the unit exceeds how many amps?
    A. 100
    B. 200
    C. 400
    D. Required for all services regardless of amperage

    Answer: _____   Reference: _____

23. Ground-fault protection is required for _____-ampere, 480-volt, 4-wire, 3-phase, solidly grounded wye services.
    A. 800
    B. 1000
    C. 1200
    D. above 200

    Answer: _____   Reference: _____

24. Impedance is present in which of the following circuits?
    A. AC only
    B. DC only
    C. Resistance only
    D. Both AC and DC

    Answer: _____   Reference: _____

25. Excluding fuses and exceptions, how many overcurrent protection devices, such as trip coils, relays, or thermal cutouts, are required on a 3-phase motor?
    A. 5
    B. 3
    C. 2
    D. 0

    Answer: _____   Reference: _____

26. The attachment plug and receptacle can be used as the controller for a portable motor that has a maximum horsepower rating of
    A. ⅓ hp
    B. ½ hp
    C. 1 hp
    D. 2 hp

    Answer: _____   Reference: _____

27. Snap switches can be grouped or ganged in outlet boxes if voltages between adjacent switches do not exceed how many volts?
    A. 200
    B. 300
    C. 400
    D. 500

    Answer: _____   Reference: _____

28. Each doorway that leads into a transformer vault from the building interior should have a tight-fitting door with a minimum fire rating of how many hours?
    A. 1
    B. 2
    C. 3
    D. 5

    Answer: _____   Reference: _____

29. The conduit run between warm and cold locations should be
    A. sealed
    B. sleeved
    C. insulated
    D. provided with drains

    Answer: _____   Reference: _____

30. Except for points of support, recessed portions of lighting fixture enclosures should be spaced at least how many inches from combustible material?
    A. ¼ inch
    B. ½ inch
    C. 1 inch
    D. 3 inches

    Answer: _____   Reference: _____

31. Surface-mounted fluorescent lighting fixtures that contain a ballast and are to be installed on combustible, low-density cellulose fiberboard should be spaced not less than how many inches from the surface of the fiberboard?
    A. ½ inch
    B. 1 inch
    C. 1½ inches
    D. 2 inches

    Answer: _____   Reference: _____

32. The minimum amount of on-site fuel for an emergency generator should be sufficient for not less than
    A. 2 hours at full demand
    B. 4 hours at 75% demand
    C. 6 hours at 50% demand
    D. 8 hours at 25% demand

    Answer: _____   Reference: _____

33. Which of the following statements about dimmers installed in theaters is (are) true?
    I.   Dimmers installed in ungrounded conductors must have overcurrent protection not greater than 125% of the dimmer rating.
    II. The circuit supplying autotransformer type dimmers must not exceed 150 volts between conductors.
    A. I only
    B. II only
    C. Both I and II
    D. Neither I nor II

    Answer: _____   Reference: _____

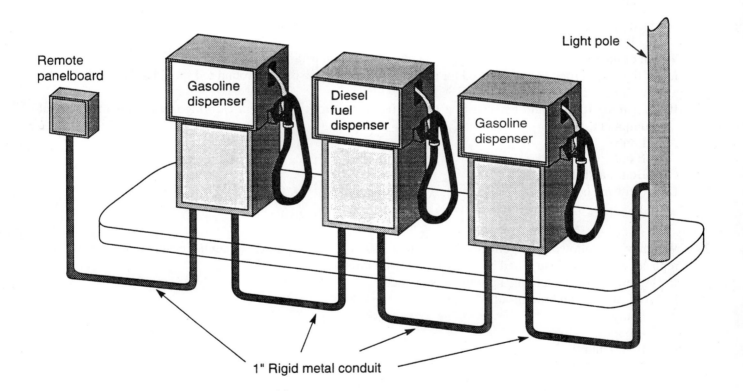

34. In the diagram above, how many conduit seals are required?
    A. 2
    B. 4
    C. 6
    D. 8

    Answer: _____   Reference: _____

35. An area must be classed as a Class II hazardous location if it contains which of the following?
    A. Combustible dust
    B. Flammable gases
    C. Ignitable liquids
    D. Ignitable vapors

    Answer:_____   Reference:_____

36. If the wiring is to be placed 18 inches above the floor, which of the following wiring methods is prohibited for a commercial automotive repair shop?
    A. Metal-clad cable
    B. Electrical metallic tubing
    C. Rigid nonmetallic conduit
    D. Nonmetallic-sheathed cable

    Answer: _____   Reference:_____

37. The minimum number of receptacles in a patient bed location of a hospital's general care area should be
    A. 1 duplex
    B. 2 duplex
    C. 3 duplex
    D. 3 single

    Answer: _____   Reference:_____

38. Which of the following statements about lightning protection is (are) true?
    I.   If lightning protection is required for an irrigation machine, a driven ground rod should be connected to the machine at a stationary point.
    II. Lightning rods should be spaced at least 6 feet away from noncurrent-carrying metal parts of electrical equipment.
    A. I only
    B. II only
    C. Both I and II
    D. Neither I nor II

    Answer: _____   Reference:_____

39. Which of the following will effectively ground a 230-volt, single-phase, residential air-conditioning unit that replaces an older unit supplied by a 2-wire, 230-volt circuit?
    I.   A new circuit containing a grounding conductor
    II. An 8-foot ground rod installed at the unit
    A. I only
    B. II only
    C. Both I and II
    D. Neither I nor II

    Answer: _____   Reference:_____

40. When connected to a made-grounding electrode, grounding electrode conductors need not be sized larger than
    A. No. 8
    B. No. 6
    C. No. 4
    D. No. 2

    Answer: _____   Reference:_____

41. Which of the following is the most acceptable grounding electrode?
    A. A driven steel approved ground rod
    B. Well casing
    C. Building foundation steel
    D. An underground cold water metal piping system

    Answer: _____   Reference: _____

42. Which of the following terms can be used in place of "resistance" in the phrase "resistance is to ohms"?
    A. Reactance
    B. Inductance
    C. Impedance
    D. Capacitance

    Answer: _____   Reference: _____

43. The *National Electrical Code*® requires a minimum of which of the following for the kitchen small-appliance load in dwelling units?
    A. Two 20 amp circuits
    B. Two 15 amp circuits
    C. One 20 amp circuit
    D. One 15 amp circuit

    Answer: _____   Reference: _____

44. Ground-fault protection of equipment is a requirement for solidly grounded wye services that are rated at more than 150 volts to ground but do not exceed 600 volts, phase-to-phase, for each service disconnecting means rated at 1,000 amps or more. The maximum amperage for this ground-fault protection relay should be set at
    A. 1,000 amps
    B. 1,200 amps
    C. 1,500 amps
    D. 2,000 amps

    Answer: _____   Reference: _____

Number	Size (kcmil)	Type	Function
2	500	THW	Phases A and B
1	400	THW	Neutral
1	300	THW	Phase C
1	No. 3	THW	Ground Conductor

45. The minimum size intermediate metal conduit (IMC) required for installation of the conductors above should not be less than
    A. 2 inches
    B. 2½ inches
    C. 3 inches
    D. 3½ inches

    Answer: _____   Reference: _____

46. The combined load of several 240-volt fixed space heaters on a 20 amp circuit should not exceed how many kilowatts?
   A. 2.4
   B. 2.6
   C. 3.8
   D. 4.8

   Answer: _____   Reference: _____

47. If three resistors with values of 5, 10, and 15 ohms, respectively, are connected in parallel, the combined resistance of the units will be
   A. 1.63 ohms
   B. 2.73 ohms
   C. 20.0 ohms
   D. 30.0 ohms

   Answer: _____   Reference: _____

48. The *National Electrical Code*® covers the installation of electrical conductors and equipment in all of the following locations except
   A. public and private buildings
   B. floating dwelling units
   C. buildings used by a utility for warehousing
   D. centers of transmission and distribution of electrical energy

   Answer: _____   Reference: _____

49. The most recent editions of the *National Electrical Code*® are revised every
   A. year
   B. 2 years
   C. 3 years
   D. 4 years

   Answer: _____   Reference: _____

50. A capacitor is located indoors. It must be enclosed in a vault if it contains more than a minimum of _____ gallon(s) of flammable liquid.
   A. 1
   B. 2
   C. 3
   D. 5

   Answer: _____   Reference: _____

Chapter Two

NATIONAL FIRE PROTECTION ASSOCIATION STANDARDS

The National Fire Protection Association (NFPA) has acted as the sponsor of the *National Electrical Code®* (*NEC®*) as well as many other safety standards. The most widely used electrical code in the world is the *NEC®*. The official designation for the *National Electrical Code®* is ANSI/NFPA 70.

NATIONAL ELECTRICAL CODE® (NEC®)

ANSI/NFPA 70 the *National Electrical Code® (NEC®)* was first developed about 100 years ago by interested industry and governmental authorities to provide a standard for the safe use of electricity. It is now revised and updated every three years. It was first developed to provide the regulations necessary for safe installations and to provide the practical safeguards of persons and property from the hazards arising from the use of electricity. It still provides these regulations and safeguards to the industry today. The *NEC®* is developed by 20 different Code Making Panels composed entirely of volunteers. These volunteers come from industry, testing laboratories, inspection agencies, engineering, users, government, and the electrical utilities. Suggestions for the content come from various sources, including individuals like you. Anyone can submit a proposal or make a comment on a proposal submitted by another individual.

For more specifics on the *NEC®* process, you can contact the National Fire Protection Association, Batterymarch Park, Quincy, MA, 02269. Request their free booklet "The NFPA Standards Making System."

The *National Electrical Code®* is advisory as far as NFPA and ANSI are concerned but is offered for use in law and regulations in the interest of life and property protection. The name *National Electrical Code®* might lead one to believe that this document is developed by the federal government. This is not so; the *NEC®* only recognizes uses of products. It approves nothing. The *NEC®* has no legal standing until it has been adopted by the *authority having jurisdiction* (see *NEC®* definition

Article 100), usually a governmental entity. Therefore, we must first check with the local electrical inspector to see which edition of the *NEC®* has been adopted. Compliance within and proper maintenance will result in an installation essentially free from hazard but not necessarily efficient, convenient, or adequate for good service or future expansion of electrical use. The *National Electrical Code®* is not an instruction manual for untrained persons nor is it a design specification. However, it does offer design guidelines.

NEC® Section 90-2 defines the scope of this document. It is intended to cover all electrical conductors and equipment within or on public and private buildings or other structures, including mobile homes, recreational vehicles, and floating buildings, and other premises such as yards, carnival, parking, and other lots, and industrial substations. Installations of conductors and equipment that connect to the supply of electricity, other installations of outside conductors on the premises, and installations of optical fiber cables are clearly covered by the *NEC®*. The *NEC®* is not intended to cover installations on ships, watercraft, railway rolling stock, automotive vehicles, underground mines, and surface mobile mining equipment. The *NEC®* is not intended to cover installations governed by the utilities, such as communication equipment, transmission, generation, and distribution installations on right-away.

> **NOTE:** For the complete list of the exemptions and coverage see 1999 *NEC®* Section 90-2. Mandatory rules are characterized by the word "Shall." Explanatory rules are in the form of fine print notes (FPN). All tables and footnotes are a part of the mandatory language. Material identified by the superscript letter "x" include text extracted from other NFPA standards and documents, such as NFPA 99 for Health Care Facilities, and NFPA 30 Flammable and Combustible Liquids Code. A complete list of all NFPA documents referenced can be found in Appendix A. New revisions are identified by a vertical line on the margin where inserted and deleted text is identified by an enlarged black dot "Bullet."

To use the *NEC*® one must first have a thorough understanding of Article 90 "The Introduction," Article 100 "Definitions," Article 110 "Requirements for Electrical Installations," and Article 300 "Wiring Methods." The rest of the Code book can be referred to on an as-needed basis.

LOCAL CODES AND REQUIREMENTS

Although most municipalities, countries, and states adopt the *National Electrical Code*®, they may not have adopted the latest edition, and in some cases the authority having jurisdiction (AHJ) may be using an older edition and not the 1999 edition. Most jurisdictions make amendments to the *NEC*® or add local requirements. These may be based on environmental conditions, fire safety concerns, or other local experience. An example of one common local amendment is that all commercial buildings be wired in metal raceways (Rigid, IMC, or EMT). The *NEC*® generally does not differentiate between wiring methods in residential, commercial, or industrial installations; however, many local jurisdictions do. Metal wiring methods, especially in fire zones, are another common amendment. Some major cities have developed their own electrical code, e.g., Los Angeles, New York, Chicago, and several metropolitan areas. In addition to the *National Electrical Code*® and all local amendments, the designer and installer must also comply with the local electrical utility rules. Most utilities have specific requirements for installing the service to the structure. There have been many unhappy designers and installers who have learned about special jurisdictional requirements after making the installation, thus incurring the costly corrections at their own expense.

If you are preparing for a state or city examination, you must check with the AHJ and see exactly what the examination content is based on. If it is a locally developed examination, the content may vary widely. If it is a nationally developed examination, it will generally be based on the latest version of the *National Electrical Code*® (1999 *NEC*®). However, some jurisdictions have supplemental examinations that cover their unique amendments, and in some jurisdictions a practical hands-on examination is given in addition to the written portion. (See Chapter One of this book for information related to some of these unique requirements.)

APPROVED TESTING LABORATORIES

Underwriters Laboratories (UL) has long been the major product testing laboratory in the electrical industry. In addition to the testing, UL is a long-time developer of product standards. By producing these standards and contracting for follow-up service after undergoing a listing procedure, a manufacturer is authorized to apply the UL label or to mark their product. In the last decade, numerous other electrical testing laboratories have arrived and are being officially recognized. Underwriters Laboratories is not the only testing laboratory evaluating electrical products. There are many testing laboratories operating today, such as Electrical Testing Laboratories (ETL), Applied Research Laboratories (APL), and Canadian Standards (CSA). Some jurisdictions evaluate and approve laboratories; others accept them based on reputation. OSHA is now evaluating and approving testing laboratories. It is the responsibility of the entity responsible for specifying the materials to verify that the product has been evaluated by a testing laboratory acceptable by the AHJ where the installation is being made. It is the installer's responsibility that the product is installed in accordance with the products listing (*NEC*® Section 110-3(b)).

CHAPTER 2 QUESTION REVIEW

> Each lesson is designed purposely to require the student to apply the entire *NEC®* text, not specific chapters or articles. It has been found that when studying for a timed, open-book examination the student must gain proficiency in the Table of Contents, the Index, and the ability to move quickly from cover to cover to find the correct answers to each question in a timely fashion.

1. Is liquidtight flexible nonmetallic conduit permitted to be used in circuits in excess of 600 volts?

   Answer: _____

   _____

   _____

   _____

   Reference: _____

2. Who is the authority having jurisdiction?

   Answer: _____

   _____

   _____

   _____

   Reference: _____

3. Name four types of installations not covered by the *NEC®*.

   Answer: _____

   _____

   _____

   _____

   Reference: _____

4. How is explanatory information characterized in the *NEC®*?

   Answer: _____

   _____

   _____

   _____

   Reference: _____

5. What section of the *NEC*® requires equipment to be installed in accordance with manufacturer's instructions?

   Answer: _____

   _____

   _____

   _____

   Reference:_____

6. When does the *NEC*® become the legal document?

   Answer: _____

   _____

   _____

   _____

   Reference:_____

7. Are all installations by utilities exempt from the *NEC*®? Explain.

   Answer: _____

   _____

   _____

   _____

   Reference:_____

8. What does the vertical line in the margin of the *NEC*® indicate?

   Answer: _____

   _____

   _____

   _____

   Reference:_____

9. What is the latest amended edition of the *NEC*®? When was the latest edition of the *NEC*® published?

   Answer: _____

   _____

   _____

   _____

   Reference:_____

10. Who set up the first meeting and when and where was it held that resulted in the *National Electrical Code®*?

   Answer: _____

   _____

   _____

   _____

   Reference: _____

11. Can Type NM cable be used for a 120/240-volt branch-circuit as temporary wiring in a building under construction where the cable is supported on insulators at intervals of not more than 10 feet?

   Answer: _____

   _____

   _____

   _____

   Reference: _____

12. Which chapter of the *NEC®* is independent of all other chapters?

   Answer: _____

   _____

   _____

   _____

   Reference: _____

13. Is the *National Electrical Code®* considered a training manual?

   Answer: _____

   _____

   _____

   _____

   Reference: _____

14. Are all mining facilities exempt from the *NEC®*?

   Answer: _____

   _____

   _____

   _____

   Reference: _____

15. A comment often made by those involved in electrical design or installation states that the *NEC*® is a true minimum. Where in the *NEC*® does it state that it is a true minimum permitted for electrical installations?

Answer: _____

_____

_____

_____

Reference: _____

16. As the inspector was making an inspection and looked at the size of overcurrent device and conductors supplying a motor, the inspector asked, "Is this motor circuit rated for continuous duty?" What is a continuous load?

Answer: _____

_____

_____

_____

Reference: _____

17. When arriving into a nearby city to make an installation, the electrical inspector informed you that they had not adopted the last two editions of the *National Electrical Code*®. Which edition would that city be enforcing?

Answer: _____

_____

_____

_____

Reference: _____

18. In wiring a small residence, the owners inform you that they wanted a doorbell mounted on the front and back of the house. What class wiring would this small 24-volt doorbell circuit be wired in and what article governs that wiring method?

Answer: _____

_____

_____

_____

Reference: _____

19. We have been asked to bid a nursing home where patients will be in varied degrees of mobility. Some will come and go freely, cook their own meals, and do their own housekeeping; others may require meals to be prepared and minor medical attention, such as someone to be sure that they take their medicine regularly. Other patients may be ambulatory and require oxygen or doctor's care. Which article of the *National Electrical Code®* covers a nursing home facility of this type?

    Answer: _____

    _____

    _____

    _____

    Reference: _____

20. In establishing a grounding electrode on a new installation, the owner says that although an underground metal water pipe exists, the owner prefers that it not be used and that a standard 8-foot ground rod be used instead. Which section of the *National Electrical Code®* governs grounding electrodes?

    Answer: _____

    _____

    _____

    _____

    Reference: _____

Chapter Three

BASIC ELECTRICAL MATHEMATICS REVIEW

In this chapter, we will check our knowledge of the basic mathematical calculations needed to perform the simple day-to-day mathematical tasks necessary to perform our work as electrical contractors or electricians. Many tasks reviewed in this chapter will appear simple. However, if you have difficulty in solving any of these problems, then it may be necessary for you to obtain a more complete study guide on basic electrical mathematics. This chapter is not a mathematical teaching aid but only a refresher to see that your mathematical skills are sufficient to solve the problems in this text.

CALCULATOR MATH

Many of us were taught and believe that mathematical study should be done manually. But in today's world, with the introduction of hand-held calculators into the industrialized world in the past 25 years, it would be somewhat silly to depend on all hand calculations. Most testing jurisdictions permit the use of hand-held, silent, battery-powered, nonprogrammable calculators to use during the test. Therefore, it is advantageous to work the problems in this chapter using the same calculator that you will take to the examination. Many calculators are on the market, priced from advertising giveaways to complex scientific notation calculators. However, for your purposes any quality hand-held calculator will do. It should have the standard engineering functions, but it is not necessary to have one with scientific capabilities. The button arrangement should be good quality. It should be a battery-powered and not a solar-powered calculator, because the room lighting might be inadequate for solar power. You should have a memory function and most modern mathematical functions, including the basic keys, input, error correction, combining operations, calculator hierarchy (calculations with a constant), roots, powers, reciprocal, factorial, percents, natural logarithm and natural antilogarithm, trigonometric functions, and error indication, accuracy and rounding, good memory usage for

store, recall and memory exchange, conversion factors from English to metric, and temperature conversions. A calculator of this quality is available in most discount stores for under $20.00. Get the calculator that you plan to use and familiarize yourself with it so you are familiar with its operations before the examination.

> **Caution:** Do not rely on programmable calculators to verify calculations or to prepare for the examination. They are unacceptable and are not permitted in most testing organizations.

Although programmable calculators are not permitted in most testing organizations, there are good programmable hand-held calculators on the market as shown in Figure 3–1 and Figure 3–2. Although these are unacceptable in an examination, they can provide a convenient tool for the contractor or electrician or those studying to improve their skills in the electrical industry.

WORKING WITH FRACTIONS

Assuming that your basic addition, subtraction, multiplication, and division skills are still sharp, we will begin with a brief review of fractions. A fraction is a part of a number. For instance, if we have a pie, we have one whole pie. If we cut that pie into four pieces, then each piece is one-fourth of a pie. The four pieces together equal four fourths, or one. If we cut that whole pie into five pieces, then each piece would be one-fifth of the pie. The five pieces together would be one pie.

If we have two pieces of a pie that has been cut into four pieces, we would have two-fourths. This fraction can be expressed in two different ways, by $\frac{2}{4}$ or $\frac{1}{2}$. To reduce a fraction is to change it into another equal fraction. To do this, you divide the numerator (top number) and the denominator (bottom number) by the same number. For instance, to reduce $\frac{2}{6}$, divide the 2 and the 6 by 2. Then the numerator would be 2 divided by 2, or 1. The denominator would be 6 divided by 2, which equals 3. Therefore, the reduced fraction is $\frac{1}{3}$; $\frac{2}{6} = \frac{1}{3}$.

Figure 3–1   An example of types of hand-held calculators suitable for quick field calculation. *(Courtesy of Calculated Industries, Inc.)*

Example: $\dfrac{2}{6} = \dfrac{2/2}{6/2} = \dfrac{1}{3}$

To reduce larger fractions, such as $^{24}\!/_{96}$ to the lowest form, we can divide both the 24 and the 96 by 24 (to get $\frac{1}{4}$) or we can divide the 24 and the 96 by 8 (which gives $^{3}\!/_{12}$) and then divide the 3 and the 12 by 3, which gives $\frac{1}{4}$, the answer.

Example: $\dfrac{24}{96} = \dfrac{24/24}{96/24} = \dfrac{1}{4}$

Reducing Mixed Numbers to Improper Fractions

In a proper fraction, the numerator is always smaller than the denominator, such as $\frac{1}{3}$, $\frac{1}{6}$, $\frac{1}{12}$. Mixed numbers are made of two numbers, a whole number and a fraction, for example: $2\frac{1}{2}$, $5\frac{1}{2}$, $3^{9}\!/_{16}$, numbers that we use every day in our work. To change these to improper fractions, the numerator will always be larger than the denominator, such as $^{12}\!/_{3}$ or $^{9}\!/_{6}$ or $^{24}\!/_{10}$. For example, to change $2\frac{1}{2}$ to an improper fraction, the 2

will be a certain number of halves $(2 = \frac{4}{2})$, which we then add to the fraction we are given $(\frac{1}{2})$. Thus, $2\frac{1}{2} = \frac{4}{2} + \frac{1}{2} = \frac{5}{2}$.

Example: $2\dfrac{1}{2} = \dfrac{2 \times 2 + 1}{2} = \dfrac{5}{2}$

To change a mixed number to an improper fraction, multiply the denominator of the fraction by the whole number and add the numerator of the fraction. Place this answer over the denominator to make an improper fraction.

Example: $5\dfrac{1}{2} = \dfrac{2 \times 5 + 1}{2} = \dfrac{11}{2}$

Changing Improper Fractions to Mixed Numbers

To change an improper fraction, such as $^{13}\!/_{3}$, to a mixed number, divide the numerator by the denominator (13 divided by 3). Any remainder is placed over the denominator (13 divided by 3 = 4 with a remainder of 1). The resulting whole number and proper

Custom LCD tells you everything you need!

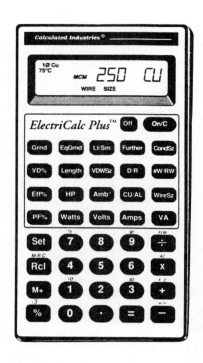

Solves electrical calculations!

Sizes wires in seconds...

...for both copper and aluminum!

Instant Conduit Sizing!

Finds new wire size to account for Voltage Drop over any distance...

...and finds the actual number or percentage of Volts dropped!

Finds Equipment Grounds!

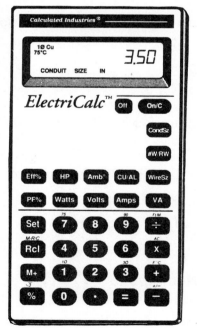

Figure 3–2 Examples of some of the types of multifunction tasks that can be quickly solved. *(Courtesy of Calculated Industries, Inc.)*

fraction form a mixed number. Thus, $^{13}\!/_3 = 4 + ^1\!/_3 = 4^1\!/_3$.

Example: $\dfrac{13}{3} = 4 + \dfrac{1}{3} = 4\dfrac{1}{3}$

Multiplication of Fractions

To multiply fractions, place the multiplication of the numerators over the multiplication of the denominators and reduce to lowest terms. For example: $^1\!/_2 \times ^3\!/_4$. Answer: $^1\!/_2 \times ^3\!/_4$ means that $1 \times 3 = 3$ and $2 \times 4 = 8$; thus, the answer is $^3\!/_8$. Example: $^1\!/_2 \times 5$. Answer: $^1\!/_2 \times 5$ means that $1 \times 5 = 5$ and $2 \times 1 = 2$, giving an improper fraction of $^5\!/_2$, which then has to be reduced to $2^1\!/_2$.

Examples: $\dfrac{1}{2} \times \dfrac{3}{4} = \dfrac{1 \times 3}{2 \times 4} = \dfrac{3}{8}$

$\dfrac{1}{2} \times 5 = \dfrac{1 \times 5}{2 \times 1} = \dfrac{5}{2} = 2\dfrac{1}{2}$

To cancel numerator and denominator means to divide both numerator and denominator by the same number. As an example, when multiplying $^3\!/_8 \times ^4\!/_9$, notice that the 3 on top and the 9 on the bottom can both be divided by 3. Cross out the 3 and write a 1 over it. Cross out the 9 and write a 3 under it. Also, the 4 on the top and the 8 on the bottom can both be divided by 4. Cross out the 4 and write a 1 over it. Cross out the 8 and write a 2 under it. The solution then is 3 reduced to 1×4 reduced to 1 (1×1) over 8 reduced to 2×9 reduced to 3 (2×3), thus giving $^1\!/_6$.

$\dfrac{3}{8} \times \dfrac{4}{9} = \dfrac{\overset{1}{\cancel{3}}}{\underset{2}{\cancel{8}}} \times \dfrac{\overset{1}{\cancel{4}}}{\underset{3}{\cancel{9}}} = \dfrac{1}{6}$

WORKING WITH DECIMALS

A decimal is a fraction in which the denominator is not written. The denominator is one. Many answers to electrical problems will be fractions, such as $^1\!/_4$ of an amp, $^1\!/_2$ of an amp, of $^1\!/_2$ of a volt, $^1\!/_3$ of a volt, etc. These will be correct mathematical answers, but they will be worthless to an electrician. The electrical measuring instruments give values expressed in decimals, not in fractions. In addition, the manufacturers of components give the values of parts in terms of decimals, such as 3.4 amps, etc., on the nameplate of a motor. Suppose we

worked out a problem and found the current in the circuit should be $^1\!/_8$ of an amp, and using the ammeter we test the circuit and find that .125 amps flow. Is our circuit correct? How will we know? How can we compare $^1\!/_8$ to .125? The easiest way is to change the fraction $^1\!/_8$ to its equivalent decimal and then compare the decimals. To change a fraction into a decimal, divide the numerator by the denominator (1 divided by 8 = .125).

$\dfrac{1}{8} = \dfrac{1}{8.000} = .125$

WORKING WITH SQUARE ROOT

Roots are the opposite of powers. A square root is the opposite of the number to the second power. The sign of the square root is $\sqrt{\quad\quad}$. To find a square root, find a number that when multiplied by itself is the number inside the square root sign. For example, to find the value of the square root of 36, ask yourself what number times itself is 36. The answer is 6, so 6 is the square root of 36.

If you can find the square root of a number with the method that uses averages, suppose that you did not know that the square root of 144 = 12. When you divide the number by its square root, the answer is the square root. If you cannot find this answer, then guess as close as possible. A good guess for 144 would be 10, because $10 \times 10 = 100$, which is close to 144. Divide 144 by 10 and the answer is 14 + a remainder. Average the guess, 10, and the answer to the division problem, 14. 10 + 14 = 24, $^{24}\!/_2 = 12$, which is the correct answer.

Follow these steps to find the square root of the larger number: Guess the answer; divide the guess into the large number; average the guess and the answer to the division problem; and check. For example, find the value of the square root of 1,024.

Step 1. Guess: in the list of square roots, $30 \times 30 = 900$. This is too small, but it is easy to divide by.

Step 2. Divide 1,024 by 30. The answer is 34, with a remainder.

Step 3. Find the average of 30 and 34. 30 + 34 = 64; $^{64}\!/_2 = 32$.

Step 4. Check: multiply 32×32. The answer is 1,024. Thus, 32 is the square root of 1,024. When you use this method to find square roots, always guess a number that ends in 0. It is easier and faster to divide by these numbers. If the average is not the square root of the number, use the average as a new guess and try again.

POWERS

A power is a product of a number multiplied by itself one or more times. For an example, what is the value of 3^2? The exponent is 2; write 3 two times, or $3^2 = 3 \times 3$; $3^2 = 9$. What is the value of 5^3? The exponent is 3; write 5 three times; in other words, $5^3 = 5 \times 5 \times 5 = 125$. These powers are terms often used in the electrical field and should be known.

OHM'S LAW

E = voltage; I = current (amperes); R = resistance

$$E = IR; \quad I = \frac{E}{R}; \quad R = \frac{E}{I}$$

As an example, all numbers are relative. For example, if E = 12, I = 3, and R = 4, then $3 \times 4 = 12$, $\frac{12}{3} = 4$, and $\frac{12}{4} = 3$. All numbers are relative. Remember, in any electrical circuit, the voltage force is the current through the conductor against its resistance. The resistance tries to stop the current from flowing. The current that flows in the circuit depends on the voltage and the resistance. The relationship among these three quantities is de-

scribed by Ohm's law. Ohm's law applies to an entire circuit or any component part of a circuit.

Voltage Drop Calculations

We have discussed the methods for finding the resistance by using Ohm's law. Now we can find the voltage drop for circuit loads using this same form.

$$E = I \times R$$

Voltage Drop Formula

Single-phase voltage drop = amperes × length × resistance × (2)

Voltage drop % = voltage drop × 100/volts

Three-phase voltage drop—calculate as a single-phase circuit and multiply resultant by .866.

$$V_{(D)} = I \times 2L \times R$$

I = Amperes
L = Length of circuit
R = Resistance values from *NEC®* Chapter 9
    Table 8
Resistance = R × [1 + a(temperature − 75)]
a = 0.00323 for copper, .000330 for aluminum
Temperature = ambient temperature

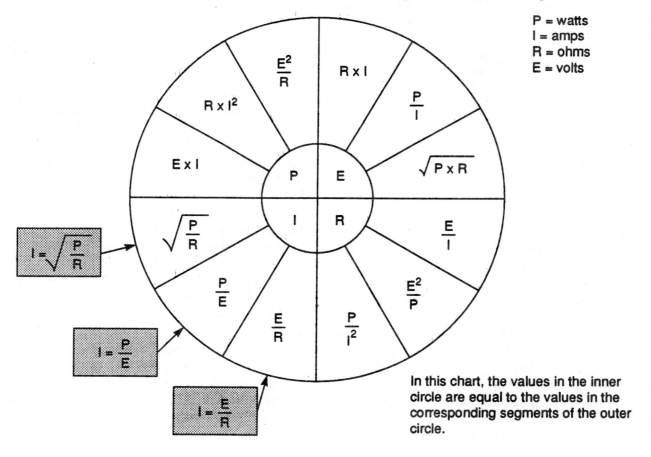

P = watts
I = amps
R = ohms
E = volts

In this chart, the values in the inner circle are equal to the values in the corresponding segments of the outer circle.

The *NEC®* references prescribe voltage drop values or percentages for feeder conductors and branch-circuit conductors; however, it should be noted these appear in a fine print note (FPN) and are only suggested and not mandatory requirements. The *NEC®* references in fine print note 2, Article 215 as not to exceed 3% of the farthest outlet of power, heating, or lighting loads or combinations of such loads, and the maximum voltage drop for both feeders and branch-circuits to the farthest outlet is not to exceed 5% and will provide reasonable efficiency of operation.

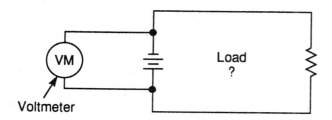

Example: A direct current circuit has 2 amperes of current flowing in it. A voltmeter reads 10 volts line to line. How much resistance is in the circuit? Answer: 5 ohms.

$$R = \frac{E}{I} \quad R = \frac{10}{2} \quad R = 5 \text{ ohms}$$

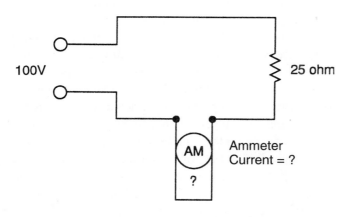

Example: If the voltage is 100 volts and the resistance is 25 ohms, what amount of current will flow in the circuit? Answer: 4 amperes.

$$I = \frac{E}{R} \quad I = \frac{100}{25} \quad I = 4 \text{ amperes}$$

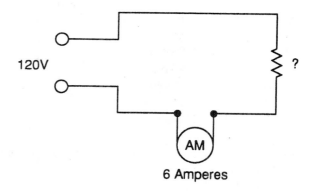

Example: If the potential across a circuit is 120 volts and the current is 6 amperes, what is the resistance? Answer: 20 ohms.

$$R = \frac{E}{I} \quad R = \frac{120}{6} \quad R = 20 \text{ ohms}$$

A series circuit may be defined as a circuit in which the resistive elements are connected in a continuous run (end to end). It is evident that because the circuit has no branches, the same current flows in each resistance. The total potential across the circuit equals the sum of potential drops across each resistance.

Example: $E_1 = IR_1$
$E_2 = IR_2$
$E_3 = IR_4$
$E = E_1 + E_2 + E_3$
$R = R_1 + R_2 + R_3$

The total potential of the circuit is:

$$E = 1R_1 + 1R_2 + 1R_3 = I(R_1 + R_2 + R_3)$$

$$I = \frac{E}{R_1 + R_2 + R_3}$$

$R_1 = 10$ ohms, $R_2 = 5$ ohms, and $R_3 = 15$ ohms.

What amount of voltage must flow to force 0.5 ampere through the circuit?

$$R_T = 5 + 10 + 15 = 30 \text{ ohms}$$

Hence, $E_T = 0.5 \times 30 = 15$ volts

What is the voltage drop across each resistance?

$E_1 = 0.5 \times 10 = 5.0$ volts
$E_2 = 0.5 \times 5 \ = 2.5$ volts
$E_3 = 0.5 \times 15 = 7.5$ volts

$E_T = 15$ volts

When the resistance and current are known, what is the formula to determine the voltage?

The current is 2 amperes, the resistance is 15 ohms, 10 ohms, and 30 ohms.

$E = IR \quad E = 2(15 + 10 + 30)$
$E = 2 \times 55 \quad E = 110$ volts

In the industry today most circuits are connected in parallel. In a parallel circuit the voltage is the same across each resistance (load).

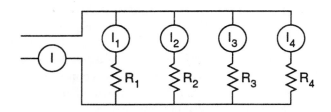

The total current is equal to the sum of all the currents.

$E = I_1R_1 = I_2R_2 = I_3R_3 = I_4R_4$ and

$I = I_1 + I_2 + I_3$

Individual Resistances =

$I_1 = \dfrac{E}{R_1} \qquad I_2 = \dfrac{E}{R_2} \qquad I_3 = \dfrac{E}{R_3} \qquad I_4 = \dfrac{E}{R_4}$

Hence: $I = \dfrac{E}{R_1} + \dfrac{E}{R_2} + \dfrac{E}{R_3} + \dfrac{E}{R_4}$

or $I = E\left(\dfrac{1}{R_1} + \dfrac{1}{R_2} + \dfrac{1}{R_3} + \dfrac{1}{R_4}\right)$

and several resistances in parallel is

$\dfrac{1}{R} = \dfrac{1}{R_1} + \dfrac{1}{R_2} + \dfrac{1}{R_3} + \dfrac{1}{R_4}$

The sum of the resistances in parallel always will be smaller than the smallest resistance in the circuit where only two resistances are in parallel.

$$R = \dfrac{R_1 \times R_2}{R_1 + R_2} \text{ ohms}$$

Power Factor

Power factor is a phase displacement of current and voltage in an AC circuit. The cosine of the phase angle of displacement is the power factor. The cosine is multiplied by 100 and is expressed as a percentage. A cosine of 90° is 0; therefore, the power factor is 0%. If the angle of displacement were 60°, the cosine of which is .5, the power factor would be 50%. This is true whether the current leads or lags the voltage. Power is expressed in DC circuits and AC circuits that are purely resistive in nature. Where these circuits contain only resistance, P (watts) = E × I. In AC circuits that contain inductive or capacity reactances, VA (volt-ampere) = E × I. In a 60 cycle AC circuit if the voltage is 120 volts, the current is 12 amperes, and the current lags the voltage by 60°, find (a) the power factor, (b) the power and voltage-amperes (VA), and (c) the power in watts. The cosine of 60 is .5; therefore, the power factor is 50%. 120 × 12 = 1,440 volt-amperes, which is called the apparent power. 120 × 12 × .5 = 720 watts, and is called the true power. Power factor is important. There is an apparent power of 1,440 VA and a true power of 720 watts. There also are 12 amperes of line current and 6 amperes of in-phase or effective current. This means that all equipment from the source of supply to the power consumption device must be capable of handling a current of 12 amperes, which actually the device is only using the current of 6 amperes. A 50% power factor was used intentionally to make the results more pronounced. The I^2R loss is based on the 12-ampere current, while only 6 amperes are effective. Power factor can be measured by a combined use of a voltmeter, ammeter, and wattmeter, or by the use of a power factor meter. When using the three meters, volt-ampere meter and wattmeter, all connected properly in the circuit, the readings of the three meters are taken simultaneously under the same load conditions and calculated as follows: power factor = true power (watts)/apparent power (which is volt-amperes). Power factor = W/EI.

CHAPTER 3 QUESTION REVIEW

Directions: Reduce the following fractions to their lowest terms.

1. $\frac{3}{6}$ = _____

2. $\frac{4}{16}$ = _____

3. $\frac{4}{40}$ = _____

4. $\frac{40}{100}$ = _____

5. $\frac{6}{8}$ = _____

6. $\frac{15}{20}$ = _____

7. $\frac{18}{36}$ = _____

8. $\frac{9}{15}$ = _____

9. $\frac{30}{50}$ = _____

10. $\frac{4}{20}$ = _____

11. $\frac{5}{20}$ = _____

12. $\frac{6}{24}$ = _____

Directions: Change the following mixed numbers to improper fractions.

1. $1\frac{1}{2}$ = _____

2. $3\frac{1}{4}$ = _____

3. $5\frac{3}{8}$ = _____

4. $4\frac{1}{4}$ = _____

5. $7\frac{1}{8}$ = _____

6. $2\frac{1}{6}$ = _____

7. $5\frac{3}{4}$ = _____

8. $8\frac{1}{3}$ = _____

9. $6\frac{4}{7}$ = _____

10. $3\frac{5}{8}$ = _____

Directions: Change the following improper fractions to mixed numbers.

1. $9/2 =$ _____

2. $12/5 =$ _____

3. $64/5 =$ _____

4. $26/8 =$ _____

5. $29/3 =$ _____

6. $31/2 =$ _____

7. $5/3 =$ _____

8. $13/3 =$ _____

Directions: Multiply the following whole numbers and fractions. Give the answer in proper reduced form.

1. $8 \times 1/2 =$ _____

2. $1/2 \times 3/4 =$ _____

3. $1/5 \times 1/8 =$ _____

4. $9 \times 1/2 =$ _____

5. $3/4 \times 7 =$ _____

6. $1/2 \times 5 =$ _____

7. $3/4 \times 12 =$ _____

8. $1/7 \times 15 =$ _____

Directions: Convert the following fractions to decimal equivalents.

1. $1/4 =$ _____

2. $5/8 =$ _____

3. $3/4 =$ _____

4. $13/15 =$ _____

5. $3/8 =$ _____

Directions: Resolve to whole numbers.

1. $2^4 =$ _____

2. $9^2 =$ _____

3. $50^2 =$ _____

4. $1^5 =$ _____

5. $10^3 =$ _____

Directions: Find the square root of the following numbers.

1. Square root of 25 = _____

2. Square root of 81 = _____

3. Square root of 49 = _____

4. Square root of 3136 = _____

5. Square root of 10201 = _____

6. Square root of 841 = _____

7. Square root of 6064 = _____

Directions: Solve the following problems.

1. A 20 amp load is fed with two conductors that have a *combined* resistance of 0.3 ohms. If the source voltage is 120 volts DC, the calculated voltage drop on this circuit is _____ %.

Answer: _____

_____

_____

_____

Reference:_____

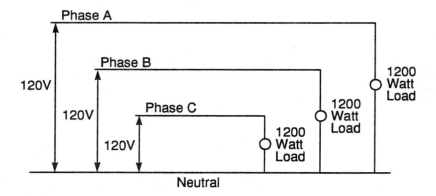

2. Refer to the above drawing.

     I(LINE A) = I(LINE B) = I(LINE C) = 10 amps.

The power factor in the circuit in the figure above is _____ .

Answer: _____

_____

_____

_____

Reference: _____

3. A 120-volt branch-circuit has only six 100-watt, 120-volt incandescent lighting fixtures connected to it. The current in the home run of this circuit is _____ amps.

Answer: _____

_____

_____

_____

Reference: _____

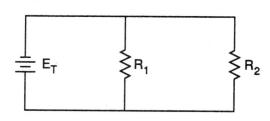

4. Refer to the above drawing.

     With a resistance at R_1 of 4 ohms and a resistance at R_2 of 2 ohms, the total resistance for the above circuit is _____ ohms.

Answer: _____

_____

_____

_____

Reference: _____

5. A 1000-watt, 120-volt lamp uses electrical energy at the same rate as a/an _____ ohm resistor.

Answer: _____

_____

_____

_____

Reference: _____

48 volts

6. Eight equal resistors are connected in series to a 48-volt source. The voltage drop across each resistor is _____ volts.

Answer: _____

_____

_____

_____

Reference: _____

This diagram applies to Questions 7 through 14.

7. What is the resistance of R_1 and R_2?

Answer: _____

_____

_____

_____

Reference: _____

8. What is the resistance of $R_3 + R_1$ and R_2?

    Answer: _____

    _____

    _____

    _____

    Reference: _____

9. What is the resistance of R_4 and R_5?

    Answer: _____

    _____

    _____

    _____

    Reference: _____

10. What is the total resistance in the circuit?

    Answer: _____

    _____

    _____

    _____

    Reference: _____

11. What is the voltage drop at R_1 and R_2?

    Answer: _____

    _____

    _____

    _____

    Reference: _____

12. What is the voltage drop at R_3?

    Answer: _____

    _____

    _____

    _____

    Reference: _____

13.   What is the voltage drop at R_4 and R_5?

Answer: _____

_____

_____

_____

Reference: _____

14.   What is the total current in this circuit?

Answer: _____

_____

_____

_____

Reference: _____

Chapter Four

INTRODUCTION TO THE *NATIONAL ELECTRICAL CODE*® (*NEC*®)

The *National Electrical Code*® is a reasonable document, developed and written in a reasonable fashion. It is a design manual, but it is not a design specification. It is not the minimum, but there are many conveniences that we take for granted that are not required by this document. There also are many safety considerations not required by this text, such as smoke detectors and other safety and detection devices that are being developed every day. The first page in *Article 90 Section 90-1(c)* says it is not an instruction manual for untrained persons. You are going to find that statement to explain much of the confusion about this document. Most engineering schools, vocational schools, and apprentice programs do not have training programs on the proper use of this document as an installation and design standard. Open your 1999 *NEC*® and look up that *NEC*® Section [90-1(c)]. You will need the 1999 *National Electrical Code*® (*NEC*®) throughout your study of this manual. There are several *NEC*® handbooks available. The NFPA *National Electrical Code*® *Handbook* contains the exact *NEC*® text in blue print and commentary in black print. If you are using that text, remember that the black commentary is only the author's opinion and may not be correct. Therefore, when studying in a classroom or for an examination, the blue text or the *NEC*® text will be that from which questions are developed.

KNOW YOUR CODE BOOK

To use the *NEC*®, we must first have a thorough understanding of:

- the Table of Contents
- the Index
- Article 90, The Introduction
- Article 100, Definitions
- Article 110, Requirements for Electrical Installations
- Article 300, Wiring Methods

We will refer to the rest of the Code book as needed; but these are general articles, and we use them continually as applicable. **You must study these articles thoroughly several times until you have a complete understanding of the content in these articles if you expect to successfully pass your examination.**

NEC® Article 90, the Introduction

Let us turn to Article 90. We often pass over the Introduction of many books that we read or study; however, the Introduction is an important part of a book. It is the application guidelines. Did you read Article 90? Read Article 90 several times. It is only about four pages, but it is a very important part of the document, especially for someone about to embark on the study of the *National Electrical Code*®.

Article 90 covers the foundation requirements of the *NEC*®. The purpose is the safeguarding of people and property from the hazards that can occur while using electricity. The Code contains rules that when complied with, and with proper maintenance normally will result in an installation that is essentially free from hazards but may not be efficient, convenient, or adequate for good service. The requirements are minimal and generally do not provide capacity for future expansion of the electrical system.

> (FPN): *Hazards often occur because of overloading of wiring systems by methods or usage not in conformity with this Code. This occurs because initial wiring did not provide for increases in the use of electricity. An initial adequate installation and reasonable provisions for system changes will provide for future increases in the use of electricity.*
>
> Reprinted with permission from NFPA 70-1999.

The *NEC*® is not a design specification, but it is a design manual. For example, specific manufacturers' products are not specified; however, no responsible designer would begin to design an electrical system without considering minimal Code requirements such as load calculations, service size and location, and grounding.

The *NEC®* is **not** an instruction manual for untrained persons.

Contained in the Introduction is the scope. It is imperative that the user know the bounds of the *NEC®*. The Code covers:

(1) Installations of electric conductors and equipment within or on public and private buildings or other structures, including mobile homes, recreational vehicles, and floating buildings; and other premises such as yards, carnival, parking, and other lots, and industrial substations.

(FPN): For additional information concerning such installations in an industrial or multibuilding complex, see the National Electrical Safety Code, *ANSI C2-1997.*

(2) Installations of conductors and equipment that connect to the supply of electricity.

(3) Installations of other outside conductors and equipment on the premises.

(4) Installations of optical fiber raceways and cable.*

Not covered by this Code is:

(1) Installations in ships, watercraft other than floating buildings, railway rolling stock, aircraft, or automotive vehicles other than mobile homes and recreational vehicles.

(2) Installations underground in mines and self-propelled mobile surface mining machinery and its attendant electrical trailing cable.

(3) Installations of railways for generation, transformation, transmission, or distribution of power used exclusively for operation of rolling stock or installations used exclusively for signaling and communications purposes.

(4) Installations of communications equipment under the exclusive control of communications utilities located outdoors or in building spaces used exclusively for such installations.

(5) Installations, including associated lighting, under the exclusive control of electric utilities for the purpose of communications, metering, generation, control, transformation, transmission, or distribution of electric energy. Such installations shall be located in buildings used exclusively by utilities for such purposes; outdoors on property

owned or leased by the utility; on or along public highways, streets, roads, etc.; or outdoors on private property by established rights such as easements.

(FPN): It is the intent of this section that this Code covers all premises wiring or wiring other than utility-owned metering equipment, on the load side of the service point of buildings, structures, or any other premises not owned or leased by the utility. Also, it is the intent that this Code cover installations in buildings used by the utility for purposes other than listed in Section 90-2(b)(5) above, such as office buildings, warehouses, garages, machine shops, and recreational buildings that are not an integral part of a generating plant, substation, or control center.*

Included in Section 90-3 is the arrangement of the Code and hierarchy. The Code is divided into an Introduction, nine Chapters, and Appendices A, B, and C.

- Chapters 1, 2, 3, and 4 apply generally and are applicable to all installations, i.e., branch circuitry, motors, appliances, and lighting.

- Chapters 5, 6, and 7 apply to special occupancies (i.e., places of assembly, hazardous locations, agriculture buildings, mobile homes, health care facilities, etc.), special equipment (i.e., swimming pools, signs, elevators, welders, etc.), or other special conditions (i.e., emergency systems, fire alarms, power limited-circuits, etc.). Although the first four still apply, these chapters augment or modify the general rules for the particular conditions.

  Example: Article 514 covers the hazardous requirements for wiring a service station. It does not cover other areas within that service station facility. It does not cover other electrical needs (i.e., service, the grounding, the branch-circuits, the feeders), and, more importantly, it does not cover other environmental concerns, such as corrosion or wet locations. The designer must take all these concerns and any others that apply and design the installation so that it will provide good service, have adequate capacity, and provide all the safety needed for the normal operation throughout the life of the facility.

- Chapter 8 is a stand-alone chapter that covers communications systems. It is independent of

* Reprinted with permission from NFPA 70-1999.

the other chapters except where they are specifically referenced.

- Chapter 9 consists of tables and examples.
- Appendix A lists extract documents. See 90-3 for an explanation.
- Appendix B is for information only and contains formula application information.
- Appendix C contains optional wire fill tables.

The Introduction lets us know that the Code is intended as suitable for mandatory application by those adopting it, i.e., governmental bodies exercising legal jurisdiction over electrical installations and for use by insurance inspectors. It also places the responsibility for enforcement on the authority having jurisdiction (AHJ).

Section 90-5 gives the style of the document and covers the mandatory rules characterized by the word *shall* and explains that explanatory material appears in the form of fine print notes, which are characterized by the designation (FPN).

Formal Interpretations

To promote uniformity of interpretation and application of the provisions of this Code, Formal Interpretation procedures have been established and may be found in the "NFPA Regulations Governing Committee Projects."

Warning: The commentary found in the NFPA *National Electrical Code Handbook is* **not** formal interpretation of the *NEC®*.

Section 90-7 states that factory-installed internal wiring or the construction of equipment need not be inspected at the time of installation of the equipment, except to detect alterations or damage, if the equipment has been listed by a qualified electrical testing laboratory that is recognized and that requires suitability for installation in accordance with this Code.

See Examination, Identification, Installation, and Use of Equipment in Section 110-3.
See definition of "Listed," Article 100.

As you can see, Article 90 is very important and one could never apply the Code correctly without first having a thorough understanding of Article 90 Introduction.

Metric Units of Measurement

For the purpose of this Code, metric units of measurement are in accordance with the modernized metric system known as the International System of Units (SI).

Values of measurement in the Code text will be followed by an approximate equivalent value in SI units. Tables will have a footnote for SI conversion units used in the table.

Conduit size, wire size, horsepower designation for motors, and trade sizes that do not reflect actual measurements, e.g., box sizes, will not be assigned dual designation SI units.

Chapter 1 "General" Requirements

You must continually refer to the definitions, *NEC®* Article 100, as you apply the requirements of the *NEC®*. These terms are unique and essential to this document, which is the reason they have been included in this document.

Article 110 covers the general requirements. Section 110-3 tells us that we must select equipment and material suitable for the installation, environment, and/or application, and that it must be installed in accordance with any instructions included with its listing or labeling. Sections 110-9 and 110-10 provide some strong mandatory requirements on interrupting rating and circuit impedance that cannot be ignored. If the equipment is intended to break current, it must be rated to interrupt the available current at fault conditions, and the system voltage and the circuit components including the impedance, shall be coordinated such that the circuit protective device will clear a fault without excessive damage. Sections 110-8 and 110-12 state that only recognized methods are permitted and that we are required to make our installation in a neat and professional manner. Section 110-14 gives mandatory requirements for making splices and terminations. Section 110-14(c) states that the terminations must be considered when determining the ampacity of a circuit. Read this *NEC®* section carefully; similar requirements have appeared in the UL Green and White books.

Section 110-12 states that we are required to carefully select the proper location for our 600 volts or less equipment (for equipment rated at over 600 volts, your code references are *NEC®* Article 110 Parts C–D, Sections 110-30 through 59) so that proper work space can be maintained. This often requires us to check the building and mechanical plans so encroachment can be avoided. To discover this encroachment in the last stages of the construction can and is often costly to one or more of the craft contractors. Other sections in Chapter 1 are equally important and must be continually referenced every time the *NEC®* is applied.

Chapter 2 Wiring and Protection

Now that we have thoroughly studied the general requirements, the next step is to study the wiring and protection requirements. Turn to Chapter 2 of the *National Electrical Code®*, Article 210, "Branch-Circuits."

> NOTE: Many requirements in this article apply only to dwellings; read the pertinent sections carefully before applying the requirements of this article.

> NOTE: Although, generally, extra circuits are not necessary, careful arrangement of the required branch-circuits is necessary for compliance with Section 210-11(a) through (c).

You also will find specific branch-circuit requirements in other articles of the *NEC®*, such as Article 430 for Motors, Article 600 for Signs, and Article 517 for Health Care Facilities. Section 210-52 gives guidelines necessary to avoid the use of extension cords in dwellings. Section 210-52(a) requires outlets to be located so that no point along the floor line is more than 6 feet from an outlet, and wall spaces 2 feet wide have an outlet in them. This requirement ensures that it is not necessary to lay cords across door openings to reach an outlet. This section also points out that outlets that are part of light fixtures and appliances, located within cabinets, or above 5½ feet cannot be counted as one of those required. Section 210-52(a) through (h) gives requirements for outlets, including the appliance outlets and the ground-fault circuit protection where required. (Several of the subsections in 210-52 reference Section 210-8 that gives the requirements for GFCIs in dwellings and other locations.) Now look at Section 210-70(a) for the required lighting outlets and wall switches. Extra lighting can be added for the customer, but the required lighting will provide a safe illumination. Section 210-70(a) generally requires a switch controlled light in each room. Stairways are required to be lighted with a switch at the top and bottom of the stairs. Section 210-70(a) now requires a switch controlled lighting outlet in attics and under floor spaces where used for storage or equipment that requires

servicing. This section also requires a switch controlled light at each outside exit or entrance. After we have established the branch-circuit requirements, we can do the general lighting load calculations in accordance with Table 220-3(b) and footnotes. The total calculated load cannot be ascertained until all of the equipment loads are known and added to the general-purpose branch-circuits. Chapter 2 Articles 215 and 225 also will guide us as we select and size inside and outside feeders.

Article 225 for outside branch-circuits and feeders has been revised in the past two *NEC®* editions, and Part B provides a mirror image of the service requirements for **"more than one building or structure on the same property under single management."** The requirements for the feeder supplying that building or structure from the service are almost identical to those for a service. Section 225-30 limits the building to be supplied by one feeder or branch-circuit except for special conditions such as fire pumps and emergency circuits and for special occupancies. These are requirements similar to those of Section 230-2. Sections 225-31 through 225-39 cover the disconnect requirements. Section 225-40 requires ready access to the overcurrent devices. Part C covers the requirements for over-600-volt feeders.

Article 230 Services is the logical next step and also the next article. We are beginning to see some rationale for the chapter and article arrangement and layout. The diagram 230-1 is an excellent aid for using this article as the sections apply to the installation. Although materials are not specified, the methods and requirements are specific. (See Figure 4–1.)

> NOTE: When designing the service or solving questions related to the service, it is important that you properly define the components of the installation (see Article 100 Definitions).

Section 230-42(a) requires the service-entrance conductors to be of sufficient size to carry the loads computed in Article 220.

Article 240 has been restructured and the scope expanded. A new Part H has been added for those portions of supervised industrial installations operating at voltages not exceeding 600 volts nominal. A new Part I has also been added for overcurrent protection over 600 volts nominal. Two major changes have been made. The overcurrent requirements for small conductors, which formerly appeared as footnotes to Table 310-16, are now appropriately located in Section 240-3(e). The requirements are now clear on how they are to be

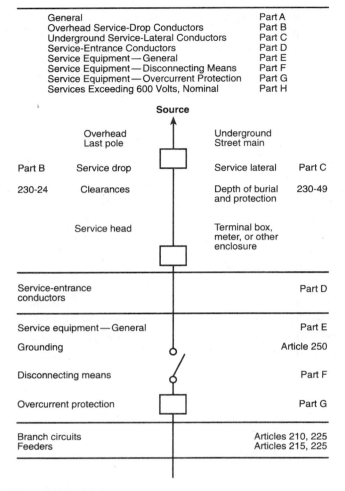

General	Part A
Overhead Service-Drop Conductors	Part B
Underground Service-Lateral Conductors	Part C
Service-Entrance Conductors	Part D
Service Equipment—General	Part E
Service Equipment—Disconnecting Means	Part F
Service Equipment—Overcurrent Protection	Part G
Services Exceeding 600 Volts, Nominal	Part H

Figure 230-1 Services.

Figure 4–1 (Reprinted with permission from NFPA 70-1999, the *National Electrical Code®*, Copyright © 1998, National Fire Protection Association, Quincy, MA 02269. This reprinted material is not the complete and official position of the National Fire Protection Association on the referenced subject, which is represented only by the standard in its entirety.)

applied, and Section 240-3(g) provides a laundry list of articles where specific overcurrent requirements are to be applied, such as for motors, Article 430.

> *(d) Small Conductors. Unless specifically permitted in (e) through (g), the overcurrent protection shall not exceed 15 amperes for No. 14, 20 amperes for No. 12, and 30 amperes for No. 10 copper; or 15 amperes for No. 12 and 25 amperes for No. 10 aluminum and copper-clad aluminum after any correction factors for ambient temperature and number of conductors have been applied.**

Section 240-3 (e) now defines tap conductors and states that they are permitted to be protected against overcurrent in accordance with Sections 210-19(d), 240-21, 364-11, 364-12, and 430-53(d).

> **Tap conductor** *is defined as a conductor, other than a service conductor, that has overcurrent protection ahead of its point of supply, that exceeds the value permitted for similar conductors that are protected as described elsewhere in this section.**

The new Part H only applies to large industrial facilities and shall only be permitted to apply to those portions of the electrical system in the supervised industrial installation used exclusively for manufacturing or process control activities.

Section 240-91 defines supervised installation as the industrial portions of a facility where all the following are met:

- Conditions of maintenance and supervision ensure that only qualified persons will monitor and service the system.
- The premises wiring system has 2,500 kVA or greater of load used in industrial process(es), manufacturing activities, or both, as calculated by Article 220.
- The premises has at least one service that is more than 150 V to ground and 300 V phase-to-phase.

This *NEC®* section makes it clear that it does not apply to buildings used as offices, warehouses, garages, etc., that are not part of the industrial plant, substation, or control center.

The new Part I has been added to include overcurrent protection for over-600-volt circuits, which previously was located in Article 710.

Article 250 Grounding. The final article in Chapter 2 is Article 250. (Article 280, Surge Arresters has not been forgotten. If you have a question related do not forget it is there.) Grounding is one important part of any project. Article 250 is specific; as you design the grounding system for this installation, you will find it to be concise and complete for most installations. However, specific grounding requirements are referenced in several other articles of the *NEC®*. If you are researching a question related to grounding, you can find the answer in the article specifically addressed in the question. (See Figure 4-2.)

---

* Reprinted with permission from NFPA 70-1999.

FPN: See Figure 250-2 for information on the organization of Article 250.

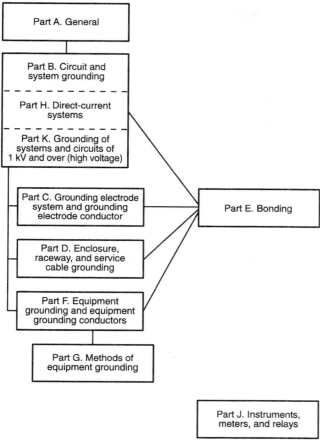

Figure 250-2   Grounding.

Figure 4–2 (Reprinted with permission from NFPA 70-1999, the *National Electrical Code®*, Copyright © 1998, National Fire Protection Association, Quincy, MA 02269. This reprinted material is not the complete and official position of the National Fire Protection Association on the referenced subject, which is represented only by the standard in its entirety.)

Article 250 has been totally reorganized for the 1999 *NEC®*. To make using this article easier, there is now a cross-reference from the 1996 *NEC®* to the 1999 *NEC®* and vice-versa as a new Appendix E. Major changes are:

- Former Section 250-51 has been relocated to Part A general Section 250-2(d). This makes it clear that in all cases the fault current path must be electrically continuous, permanent, and capable of carrying the maximum fault likely to be imposed on it. It must have sufficiently low impedance to facilitate the operation of the overcurrent device(s) under fault conditions.

- New Section 250-32, formerly Section 250-24, has been revised to make it clear that the neutral shall not be re-grounded at the second building or structure if there are any common metallic paths, such as water line, gas line, telephone, CATV, or process piping. This means that in most cases, an equipment grounding conductor (metal raceway) must be run with the circuit conductors.

- Former Section 250-81(a) has been relocated from Part H to Part C and now requires that the supplemental electrode comply with Section 250-56 (old Section 250-84). This means that unless you test the resistance of the ground rod and it has 25 ohms or less, you must drive two electrodes to supplement the water pipe.

Now that we have our branch-circuitry, feeders, service, and grounding designed into the installation, it is time to select the wiring methods and materials, which are found in Chapter 3 of the *NEC®*.

Chapter 3 Wiring Methods. As stated earlier, this article covers general requirements that must be applied to all wiring methods. The wiring method selected must be appropriate for the environment, meet the physical protection requirements, temperature, and all other pertinent conditions that can affect the material selected. The condition or question can address the wiring method directly or can require you to select an approved method based on these factors. Be sure to study all of the "uses permitted" and the "uses not permitted" in each specific article carefully.

For example, when selecting rigid metal conduit, you need to consider only the corrosion elements. However, when selecting nonmetallic sheath cable (Romex), many factors must be considered. Once this has been done, a wiring method can be selected from Chapter 3 based on these factors.

Temporary Wiring. Article 305 permits the use of a lower class wiring for any temporary wiring needed during the construction phase.

The conductor type is selected and sized from Article 310. The wiring method we select may need considering at the same time as the conductor type. Most installations require several wiring methods to meet all conditions, such as a method that is acceptable for dry concealed locations. For this location most methods in Chapter 3 are acceptable. We also may need to select a flexible wiring method for equipment connection and a material that can be used in wet locations and underground. There are methods acceptable in all locations;

all are found in Chapter 3 including the junction, outlet, cabinets, and cutout boxes from Articles 370 and 373 (to identify these, check Article 100 Definitions). Again, you are required to comply with all applicable provisions in Article 300 and in that article covering the specific wiring method.

The requirements for switches (Article 380), switchboards and panelboards (Article 384) are also covered in Chapter 3 of the *NEC®*. Read the scope of these articles. You see that Article 380 applies to all switches, switching devices, and circuit breakers where used as switches. In Article 384 the scope is not as clear, and this is an example in which the *NEC®* can be difficult to interpret. Motor control centers are not mentioned in Article 384, yet fine print note (FPN) to *NEC®* Section 430-1 states: "Installation requirements for motor control centers are covered in Section 110-26(f)." You must study the code carefully and thoroughly to successfully pass the examination the first time. Answer the following Question Review carefully. You will find it necessary to refer to each article in the *National Electrical Code®*, the Table of Contents, and the Index to solve these questions. You will find that answering these questions will prepare you for latter chapters of this book. The next eight chapters of the manual will help you gain needed understanding of the *NEC®* and give you various sample questions on each chapter of the *NEC®*.

CHAPTER 4 QUESTION REVIEW

Answer the following Question Review carefully. You will find it necessary to refer to each article in the *National Electrical Code®*, the Table of Contents, and the Index to answer these questions. You will find that answering these questions will prepare you as you proceed through the latter chapters of this book. The next eight chapters of the manual will help you gain that needed understanding of the *NEC®* and give you a wide variety of sample questions on each chapter of the *NEC®*.

> **Each lesson is designed purposely to require the student to apply the entire *NEC®* text, not specific chapters or articles. It has been found that when studying for a timed, open-book examination the student must gain proficiency in the Table of Contents, the Index, and the ability to move quickly from cover to cover to find the correct answers to each question in a timely fashion.**

1. Would a floating restaurant located in a river or along a dock in a harbor be covered by the *National Electrical Code®*?

   Answer: _____

   _____

   _____

   _____

   Reference: _____

2. In which article in the *NEC*® would you find the requirements for grounding a community antenna television and radio distribution system?

Answer: _____

_____

_____

_____

Reference: _____

3. In which article in the *NEC*® would you find the requirements for receptacles, cord connectors, and attachment plugs?

Answer: _____

_____

_____

_____

Reference: _____

4. All requirements for the installation and calculation of branch-circuits are found in Chapter 2 of the *NEC*®. (True or False)

Answer: _____

_____

_____

_____

Reference: _____

5. In what article would the requirements for installing the light fixtures in a paint spray booth be found?

Answer: _____

_____

_____

_____

Reference: _____

6. In what article would you find the requirements for the installation of central heating equipment, such as a natural gas furnace, be found?

Answer: _____

_____

_____

_____

Reference: _____

7. In what section of the *National Electrical Code*® would you find the allowable number of conductors in a conduit?

   Answer: _____

   _____

   _____

   _____

   Reference: _____

8. Which article of the *NEC*® covers the power and lighting in mine shafts?

   Answer: _____

   _____

   _____

   _____

   Reference: _____

9. In which *NEC*® article and section would you find the cover requirements for underground conductors running from building A to building B? (The conductors are direct burial conductors, the circuit is a 3-phase, 2,300-volt, 4-wire system.)

   Answer: _____

   _____

   _____

   _____

   Reference: _____

10. The conductors in Question 9 above must terminate in a disconnecting means. What article and section of the *NEC*® cover the requirements for the disconnecting equipment at building B?

    Answer: _____

    _____

    _____

    _____

    Reference: _____

11. A 4-inch rigid nonmetallic conduit is being installed on the outside of a building. The length of the conduit from the junction box to the point that it enters the building is 300 feet. Where in the *National Electrical Code*® is the amount of the expansion of the rigid nonmetallic conduit found?

Answer: _____

_____

_____

_____

Reference: _____

12. You are going to wire a nightclub that will seat 300 people. Which article in the *NEC*® covers the wiring requirements for this nightclub?

Answer: _____

_____

_____

_____

Reference: _____

13. In which article and section in the *National Electrical Code*® would the requirements for grounding a portable generator be found?

Answer: _____

_____

_____

_____

Reference: _____

14. In which article and section in the *NEC*® would the requirements for bonding the forming shell of an inground swimming pool be found?

Answer: _____

_____

_____

_____

Reference: _____

15. In which article and section in the *NEC*® is the meaning of the small superscript x explained?

Answer: _____

_____

_____

_____

Reference: _____

16. Where in the *NEC*® are the requirements for the location and installation of smoke detectors found?

Answer: _____

_____

_____

_____

Reference: _____

17. Under what article of the *National Electrical Code*® would the wiring requirements for fire alarms be found?

Answer: _____

_____

_____

_____

Reference: _____

18. An existing building is being remodelled. Permanent receptacles in the building are being used to supply temporary power for the construction. Must these permanent receptacles be GFCI protected?

Answer: _____

_____

_____

_____

Reference: _____

19. You have been given the responsibility for wiring a phosphoric acid fertilizer facility. What section of the *National Electrical Code*® covers this type of hazardous installation: (a) What class wiring is required? (b) If the hazardous dusts are not controlled and will be found on the day-to-day work schedule, what division must this area be wired in?

Answers: _____

_____

_____

_____

Reference: _____

20. You have been asked to wire a neon sign in the bedroom of a new home you are wiring. Which section of the *National Electrical Code*® would you wire this neon sign for the bedroom?

Answer: _____

_____

_____

_____

Reference:_____

Chapter Five

GENERAL WIRING REQUIREMENTS

The general requirements for the *National Electrical Code*® are found in Chapter 1. They include Article 100, which are the definitions, Part A—General Definitions, and Part B—the definitions for over 600 volts nominal. Article 110 gives the requirements for electrical installations, Part A—General Requirements, and Part B—the requirements for over 600 volts nominal. Article 100 contains the definitions essential to the proper application of the *National Electrical Code*®. They do not include commonly defined general terms or commonly defined technical terms from other related codes and standards. In general, only the terms used in two or more articles are defined in Article 100. Other definitions are included in the articles for which they are used but may be referenced in Article 100. Part A of Article 100 contains definitions intended to apply wherever the terms are used throughout the *NEC*®. Part B contains only the definitions applicable to installations and equipment operating at over 600 volts nominal. Electrical terms not specific to the *NEC*® may be defined in the *IEEE Dictionary*. Other words are defined in *Webster's Dictionary*. An example of a definition with a specific meaning in the *NEC*® found in Article 100 would be **"accessible."** Accessible is defined both for wiring methods and equipment. The definition found in Article 100 applies specifically to its use throughout the *NEC*®. Other terms or words found in Article 100 meanings generally differ from *Webster's Dictionary*.

Article 110 contains the general requirements for all electrical installations. It contains the approval required for installations and equipment (see Figure 5–1) and instructions on the examination of the equipment to judge it suitable. It includes information such as the voltages considered, conductor information, and many terms to describe the uses throughout the Code. Section 110-8 reminds us that only wiring methods recognized as suitable are included in the Code, the recognized methods of wiring permitted to be installed in any type of building or occupancy except as otherwise provided in the *NEC*®. This is important because often materials are misused and wiring methods are devised in the

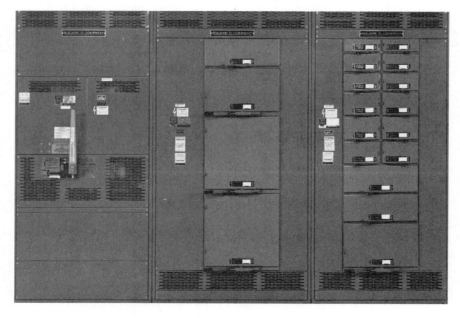

Figure 5–1 Electrical switchboard. *(Courtesy of Square D Company)*

field that are not in accordance with the manu- facturer's instructions or design recommendations of the equipment being used. General requirements remind us that all circuitry and equipment intended to break current must be properly sized and installed so proper operation occurs. Article 110 reminds us that the proper equipment must be used where chemicals, gases, vapors, fumes, liquids, or other agents having a deteriorating effect on equipment must be considered. It reminds us that all work must be installed in a neat and workmanlike manner and that unused openings must be closed.

Wire Temperature Termination Requirements

Wire temperature ratings and temperature termination requirements for equipment result in rejected installations. Information about this topic can be found in testing agency directories, product testing standards, and manufacturers' literature, but most do not consult these sources until it is too late.

A section in the 1999 *National Electrical Code®*, Section 110-14(c), provides the information contained in the standards to the inspectors, installers, and engineers.

Why Are Temperature Ratings Important?

Conductors carry a specific temperature rating based on the type of insulation employed on the conductor. Common insulation types can be found in Table 310-13 of the *NEC®*, and corresponding ampacities can be found in Table 310-16.

Example: A 1/0 copper conductor ampacity based on different conductor insulation types:

Insulation Type	Temperature Rating	Ampacity
TW	60°C	125 amperes
THW	75°C	150 amperes
THHN	90°C	170 amperes

Although the wire size has not changed (1/0 Cu), the ampacity *has* changed due to the temperature rating of the insulation on the conductor. Higher-rated insulation allows a smaller conductor to be used at the same ampacity as a larger conductor with lower-rated insulation, and, as a result, the amount of copper and even the number of conduit runs needed for the job may be reduced.

One common misapplication of conductor temperature ratings occurs when the rating of the equipment termination is ignored. Conductors must be sized by considering where they will terminate and how that termination is rated. If a termination is rated for 75°C, this means the temperature at that termination may rise to 75°C when the equipment is loaded to its ampacity. If 60°C insulated conductors were employed in this example, the additional heat at the connection above the 60°C conductor insulation rating could result in failure of the conductor insulation.

When a conductor is selected to carry a specific load, the user/installer or designer must know termination ratings for the equipment involved in the circuit.

Example: Using a circuit breaker with 75°C terminations and a 150-ampere load, if a THHN (90°C) conductor is selected for the job, from Table 310-16 select a conductor that will carry the 150 amperes. Although Type THHN has a 90°C ampacity rating, the ampacity from the 75°C column must be selected because the circuit breaker termination is rated at 75°C. Looking at the table, a 1/0 copper conductor is acceptable. The installation would be as shown in Figure 5–2, with proper heat dissipation at the termination and along the conductor length. If the temperature rating of the termination had not been considered, a No. 1 AWG conductor based on the 90°C ampacity may have been selected, which may have led to overheating at the termination or premature opening of the over-

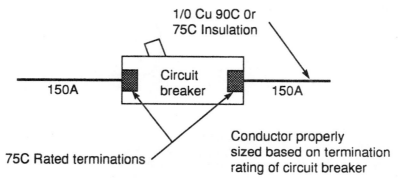

Figure 5–2 *(Courtesy of Square D Company)*

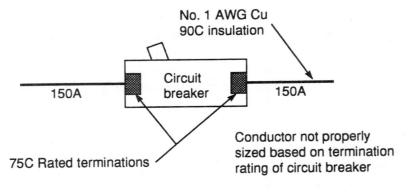

Figure 5–3   *(Courtesy of Square D Company)*

current device because of the smaller conductor size. (See Figure 5–3.)

In the example above, a conductor with a 75°C insulation type (THW, RHW, USE, and so forth) also would be acceptable because the termination is rated at 75°C. A 60°C insulation type (TW or perhaps UF) is not acceptable because the temperature at the termination could rise to a value greater than the insulation rating.

General Rules for Application

When applying equipment with conductor terminations, the following two basic rules apply:

- Rule 1 [*NEC*® Section 110-14(c)(1)]—Termination provisions for equipment rated 100A or less or equipment that is marked for No. 14 to No. 1 AWG conductors are rated for use with conductors rated 60°C. (See Figure 5–4.)

- Rule 2 [*NEC*® Section 110-14(c)(2)]—Termination provisions for equipment rated greater than 100A or equipment terminations marked for conductors larger than No. 1 AWG are for use with conductors rated 75°C. (See Figure 5–5.)

There are exceptions to the rules. Two important exceptions are:

- Exception 1—Conductors with higher temperature insulation can be terminated on lower temperature rated terminations provided the ampacity of the conductor is based on the lower rating. This is illustrated in the example where the THHN (90°C) conductor ampacity is based on the 75°C rating to terminate in a 75°C termination. The following table provides a quick reference of how this exception would apply to common terminations.

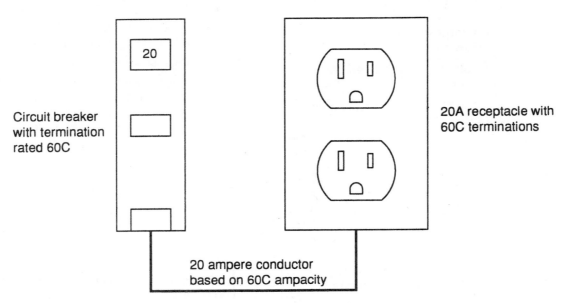

Figure 5–4   *(Courtesy of Square D Company)*

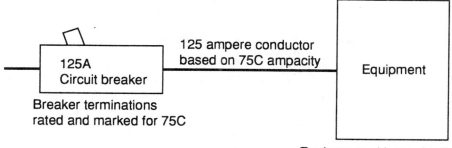

Figure 5–5 *(Courtesy of Square D Company)*

Conductor Insulation Versus Equipment Termination Ratings

Termination Rating	60°C	Conductor Insulation Rating 75°C	90°C
60°C	OK	OK (at 60°C ampacity)	OK (at 60°C ampacity)
75°C	No	OK	OK (at 75°C ampacity)
60/75°C	OK	OK (at 60°C or 75°C ampacity)	OK (at 60°C or 75°C ampacity)
90°C	No	No	OK*

* The equipment must have a 90°C rating to terminate 90°C wire at its 90°C ampacity.

- Exception 2—The termination can be rated for a value higher than the value permitted in the general rules if the equipment is listed and marked for the higher temperature rating.

Example: A 30-ampere safety switch could have 75°C rated terminations if the equipment was listed and identified for use at this rating.

A Word of Caution

When terminations are inside equipment, such as panelboards, motor control centers, switchboards, enclosed circuit breakers, and safety switches, it is important to note that the temperature rating identified on the equipment labeling should be followed—not the rating of the lug. It is common to use 90°C rated lugs (i.e., marked AL9CU), but the equipment rating may be only 60°C or 75°C. The use of the 90°C rated lugs in this type of equipment does not give permission for the installer to use 90°C wire at the 90°C ampacity.

The labeling of all devices and equipment should be reviewed for installation guidelines and possible restrictions.

Equipment Terminations Available Today

Remember, a conductor has two ends, and the termination on each end must be considered when applying the sizing rules.

Example: Consider a conductor that will terminate in a 75°C rated termination on a circuit breaker at one end and a 60°C rated termination on a receptacle at the other end. This circuit must be wired with a conductor that has an insulation rating of at least 75°C (due to the circuit breaker) and sized based on the ampacity of 60°C (due to the receptacle).

In electrical equipment, terminations are typically rated at 60°C, 75°C, or 60/75°C. There is no listed distribution or utilization equipment that is listed and identified for the use of 90°C wire at its 90°C ampacity. This includes distribution equipment, wiring devices, transformers motor control devices, and utilization equipment such as HVAC, motors, and light fixtures. Installers and designers who do not know this fact have been faced with jobs that do not comply with the

National Electrical Code®, and jobs have been turned down by the electrical inspectors.

Example: A 90°C wire can be used at its 90°C ampacity (see Figure 5–6). Note that the No. 2 90°C rated conductor does not terminate directly in the distribution equipment but in a terminal or tap box with 90°C rated terminations.

Frequently, manufacturers are asked when distribution equipment will be available with terminations that will permit 90°C conductors at the 90°C ampacity. The answer is complex and requires not only significant equipment redesign (to handle the additional heat) but also coordination of the downstream equipment where the other end of the conductor will terminate. Significant changes in the product testing/listing standards also would have to occur.

A final note about equipment is that, generally, equipment requiring the conductors to be terminated in the equiment has an insulation rating of 90°C but has an ampacity based on 75°C or 60°C. This type of equipment might include 100% rated circuit breakers, fluorescent lighting fixtures, and so on, and will include a marking to indicate such a requirement. Check with the equipment manufacturer to see if any special considerations need to be considered.

Higher-Rated Conductors and Derating Factors

One advantage to conductors with higher insulation ratings is noted when derating factors are applied. Derating factors can be required because of the number of conductors in a conduit, higher ambient temperatures, or possibly internal design requirements for a facility. By beginning the derating process at the conductor ampacity based on the higher insulation value, upsizing the conductors to compensate for the derating may not be required.

For the following example of this derating process, the following two points must be considered:

1. The ampacity value determined after applying the derating factors must be equal to or less than the ampacity of the conductor, based on the temperature limitations at its terminations.

2. The derated ampacity becomes the allowable ampacity of the conductor, and the conductor must be protected against overcurrent in accordance with this allowable ampacity.

Example for the derating process: Assume a 480Y/277 VAC, 3∅4W feeder circuit to a panelboard supplying 200 amperes of fluorescent

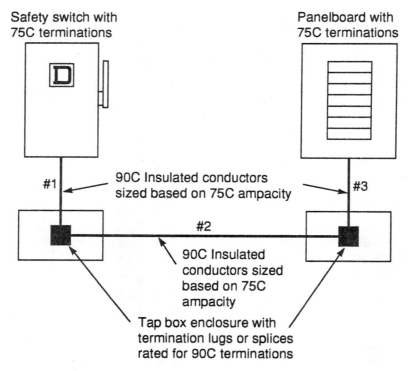

Safety switch with
75C terminations

Panelboard with
75C terminations

#1

90C Insulated conductors
sized based on 75C ampacity

#3

#2

90C Insulated
conductors sized
based on 75C
ampacity

Tap box enclosure with
termination lugs or splices
rated for 90C terminations

Figure 5–6   *(Courtesy of Square D Company)*

lighting load, and assume that the conductors will be in a 40°C ambient temperature. Also, assume that the conductors originate and terminate in equipment with 75°C terminations.

1. Because the phase and neutral conductors will be in the same conduit, the issue of conduit fill must be considered. Table 310-15(6)(4) states that the neutral must be considered to be a current-carrying conductor because it is supplying electric discharge lighting.
2. With four current-carrying conductors in the raceway, apply Table 310-15(6)(2). This requires an 80% reduction in the conductor ampacity based on four to six current-carrying conductors in the raceway.
3. The correction factors at the bottom of Table 310-16 also must be applied. An adjustment of .88 for 75°C and .91 for 90°C is required where applicable.

Now the calculations:

Using a 75°C conductor such as THWN:

300 kcmil copper has a 75°C ampacity of 285 amperes. Using the factors from above, the calculations are:

$$285 \times .80 \times .88 = 201 \text{ amperes}$$

Thus, 201 amperes is now the allowable ampacity of the 300 kcmil copper conductor for this circuit. If the derating factors for conduit fill and ambient temperatures had not been required, a 3/0 copper conductor would have met these requirements.

Using a 90°C conductor such as THHN:

250 kcmil copper has a 90°C ampacity of 290 amperes. Using the factors from above, the calculations are:

$$290 \times .8 \times .91 = 211 \text{ amperes}$$

211 amperes is less than the 75°C ampacity of a 250 kcmil copper conductor (255 amperes), so the 211 amperes would now be the allowable ampacity of the 250 kcmil conductor. If the calculation resulted in a number larger than the 75°C ampacity, the actual 75°C ampacity would have been required to be used as the allowable ampacity of the conductor. This is critical because the terminations are rated at 75°C.

NOTE: The primary advantage to using 90°C conductors is exemplified by this example. The conductor is permitted to be reduced by one size (300 kcmil to 250 kcmil) and still accommodate all required derating factors for the circuit.

In summary, when using 90°C wire for derating purposes, begin by derating at the 90°C ampacity. Compare the result of the calculation with the ampacity of the conductor based on the termination rating (60°C or 75°C). The smaller of the two numbers then becomes the allowable ampacity of the conductor. Note that if the load dictates the size of the required overcurrent device (i.e., continuous load × 125%), then the conductor allowable ampacity must be protected by the required overcurrent device. (See Section 240-3 that permits the conductor to be protected by the next higher standard size overcurrent device as per Section 240-6.) This may require moving to a larger conductor size and beginning the derating. In the example, if the 200 amperes of load were continuous, a 250-ampere overcurrent device would be required. The 201 or 211 amperes from the calculations would not be protected by the 250-ampere overcurrent device and, as such, we would be required to move to a larger conductor meeting these requirements to begin the derating process.

Summary

Several factors affect how the allowable ampacity of a conductor is determined. The key is not to treat the wire as a system but as a component of the total electrical system. The terminations, equipment ratings, and environment affect the ampacity assigned to the conductor. If the designer and the installer remember each rule, the installation will go smoother.

Section 110-14 is especially important in that the requirements for all electrical connections must be made in the proper manner. Section 110-16 reminds us that we must have sufficient working space around electrical equipment of 600 volts nominal or less; that we must have clearances in front and above (again see chapter opener photo of switchboard, Figure 5–1); we must have headroom to work on the equipment; proper illumination; and that all live parts must be guarded against accidental contact. Article 110 reminds us that warning signs must be placed and that electrical equipment must be protected from physical damage. Section 110-22 states that each disconnecting means for motors and appliances, and service, feeders, or branch-circuits at the point where it originates must be legibly identified to indicate its purpose and that the identifying markings must be sufficiently durable to withstand the environment involved. Where circuits and fuses are installed with a series combination rating "Caution—Series Rated System," the equipment enclosure shall be legibly marked in the field to indicate the equipment has been applied with a series combination rating. Over-600-volt requirements are found in Part B of Article 110. These requirements are specific about entrance and access to working space, guarding barriers, enclosures for electrical installations (see Figure 5–1), and separation from other circuitry. Over-600-volt installations must be located in locked rooms or enclosures, except where under the observation of qualified persons at all times. Table 110-34(e) covers the elevation of unguarded live parts above working space for nominal voltages 1001 through 35 kV.

Working Clearances

Section 110-2 reminds us that we must have sufficient working space around electrical equipment of 600 volts nominal or less, that we must have clearances in front and above (see switchboard Figure 5–1), and that all live parts must be guarded against accidental contact. Article 110-2 reminds us that warning signs must be placed and that electrical equipment must be protected from physical damage. (See Figure 5–7.)

Section 110-2(e) states that a minimum of 6½ feet or to the top of the equipment, whichever is greater, must be maintained in the required working spaces to work on the equipment. This means that the area 30 inches wide or the width of the equipment, whichever is greater, by the depth required in Table 110-26(a) must have a minimum head room clearance of the height of the equipment or 6½ feet, whichever is greater, to provide a safe place for the worker to work on the equipment. Proper illumination must be provided in these working spaces and about the equipment. If an adequate lighting source is provided in these rooms or areas, additional specified lighting is not required.

Section 110-22 states that each disconnecting means for motors and appliances, and service, feeders, or branch-circuits at the point where it originates must be legibly identified to indicate its purpose. Also, the identifying markings must be sufficiently durable to with-

---

Section 110-14(c) **Temperature Limitations.** *The temperature rating associated with the ampacity of a conductor shall be so selected and coordinated so as not to exceed the lowest temperature rating of any connected termination, conductor, or device. Conductors with temperature ratings higher than specified for terminations shall be permitted to be used for ampacity adjustment, correction, or both.*

(1) Termination provisions of equipment for circuits rated 100 amperes or less, or marked for Nos. 14 through 1 conductors, shall be used only for one of the following:

(a) *Conductors rated 60°C (140°F), or*
(b) *Conductors with higher temperature ratings, provided the ampacity of such conductors is determined based on the 60°C (140°F) ampacity of the conductor size used, or*
(c) *Conductors with higher temperature ratings if the equipment is listed and identified for use with such conductors, or*
(d) *For motors marked with design letters B, C, D, or E, conductors having an insulation rating of 75°C (167°F) or higher shall be permitted to be used provided the ampacity of such conductors does not exceed the 75°C (167°F) ampacity.*

(2) Termination provisions of equipment for circuits rated over 100 amperes, or marked for conductors larger than No. 1, shall be used only for

(a) *Conductors rated 75°C (167°F), or*
(b) *Conductors with higher temperature ratings provided the ampacity of such conductors does not exceed the 75°C (167°F) ampacity of the conductor size used, or up to their ampacity if the equipment is listed and identified for use with such conductors.*

(3) Separately installed pressure connectors shall be used with conductors at the ampacities not exceeding the ampacity at the listed and identified temperature rating of the connector.

FPN: With respect to Sections 110-14(c)(1), (2), and (3), equipment markings or listing information may additionally restrict the sizing and temperature ratings of connected conductors.

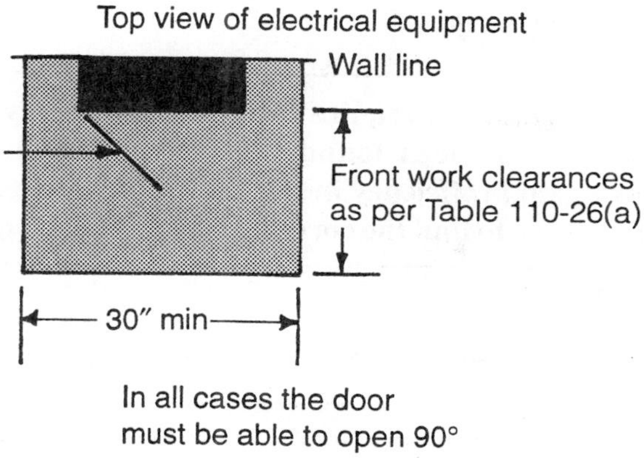

Top view of electrical equipment

Wall line

Front work clearances
as per Table 110-26(a)

|←——— 30″ min———→|

In all cases the door
must be able to open 90°

Figure 5–7 Section 110-26(a) requires that a minimum of clearance of 30 inches wide be maintained in front of electrical equipment or the width of the equipment, whichever is greater; in addition, all doors must have clearance to open at least 90°. The distance in front of the equipment must be in accordance with Table 110-26(a). See Part C to Article 110 for required clearances over 600 volts and Table 110-34(a) for clearances.

stand the environment involved. Where circuits and fuses are installed with a series combination rating "Caution—Series Rated System," the equipment enclosure shall be legibly marked in the field to indicate the equipment has been applied with a series combination rating.

Working Clearances Over 600 Volts

Over-600-volt requirements that supplement or modify the preceding sections are found in Part B. In no case shall the provisions of this part apply to equipment on the supply side of the service conductors. These requirements are very specific regarding entrance and access to working space, guarding barriers, enclosures

for electrical installations, and separation from other circuitry. (See Figure 5–1.)

Over-600-volt installations must be located in locked rooms or enclosures, except where they are under the observation of qualified persons at all times. Table 110-34(e) covers the elevation of unguarded live parts above working space for nominal voltages 1001 through 35 kV.

Many of the requirements for over-600-volt installations are similar to those for 600 volts or less; however, when preparing for an examination or applying these sections, it is very important that these sections be studied carefully.

CHAPTER 5 QUESTION REVIEW

> Each lesson is designed purposely to require the student to apply the entire *NEC®* text, not specific chapters or articles. It has been found that when studying for a timed, open-book examination the student must gain proficiency in the Table of Contents, the Index, and the ability to move quickly from cover to cover to find the correct answers to each question in a timely fashion.

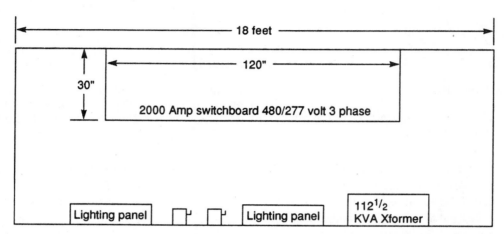

This proposed electrical equipment room is to be constructed of concrete floor and concrete. Block walls will be painted.

1. In the example above, what is the space required between the switchboard and the 112½ kVA transformer?

   Answer: _____

   _____

   _____

   _____

   Reference: _____

2. In the example above, are one or two doors required for entering and leaving this room?

   Answer: _____

   _____

   _____

   _____

   Reference: _____

3. Is the electrical equipment room in the example above required to be illuminated?

   Answer: _____

   _____

   _____

   _____

   Reference: _____

4. What is the required headroom in front of the electrical equipment needed for those performing tests or maintenance on this electrical equipment?

Answer: _____

_____

_____

_____

Reference: _____

5. In the example above, if the proposed electrical room size was increased from 10×18 to 18×18, how many doors would be required, provided that the room had no additional electrical equipment added?

Answer: _____

_____

_____

Reference: _____

Heat pump

6. Does the installation as shown above comply with the *NEC®*? It is a single-family dwelling. The supplementary overcurrent device and required disconnecting means are located on the dwelling wall directly behind the heat pump for convenience.

Answer: _____

_____

_____

_____

Reference: _____

7. The working space in front of enclosed electrical equipment operating at 480 volts must have a width when facing the equipment of at least _____ inches.

   Answer: _____

   _____

   _____

   _____

   Reference: _____

8. In an electrical system, an overcurrent device shall be placed in series with each _____ conductor.

   Answer: _____

   _____

   _____

   _____

   Reference: _____

9. A conduit nipple of 18 inches in length is installed between two boxes. This nipple can be filled to a maximum of _____ % without derating the conductors.

   Answer: _____

   _____

   _____

   _____

   Reference: _____

10. A distribution panelboard must not be installed in a _____.

    Answer: _____

    _____

    _____

    _____

    Reference: _____

11. An unprotected cable is installed through bored holes in wood studs. The holes must be bored so the edge of the hole is at least _____ inch(es) from the nearest edge of the stud.

    Answer: _____

    _____

    _____

    _____

    Reference: _____

12. Circuit conductors run in electrical nonmetallic tubing (ENT) cannot exceed _____ volts.

Answer: _____

_____

_____

_____

Reference: _____

13. Can liquidtight flexible nonmetallic conduit be used as a service raceway?

Answer: _____

_____

_____

_____

Reference: _____

14. When a metal raceway is used as physical protection for a grounding electrode conductor, must this metallic raceway be bonded even though the grounding-electrode conductor is bare copper? If the answer is yes, how must this metal raceway be bonded?

Answer: _____

_____

_____

_____

Reference: _____

15. The size of the equipment grounding conductor routed with the feeder or branch-circuit conductors is determined by what?

Answer: _____

_____

_____

_____

Reference: _____

16. How many outdoor receptacle outlets are required for a one-family dwelling?

Answer: _____

_____

_____

_____

Reference: _____

17. Are the metal parts of electrical equipment associated with a hydromassage tub required to be bonded?

    Answer: _____

    _____

    _____

    _____

    Reference: _____

18. Where rooms within dwellings are separated by railings, planters, and so forth, are receptacles required to be installed in or along these items?

    Answer: _____

    _____

    _____

    _____

    Reference: _____

19. Are the receptacle outlet spacings for room dividers, bar-type counters, and fixed room dividers the same for site-built dwellings as they are for mobile homes?

    Answer: _____

    _____

    _____

    _____

    Reference: _____

20. You have been asked to install a receptacle on a rooftop for the air-conditioning repair people to plug their instruments and portable tools into. In which section of the *National Electrical Code*® would this be found?

    Answer: _____

    _____

    _____

    _____

    Reference: _____

Chapter Six

BRANCH-CIRCUITS AND FEEDERS

Branch-circuits are covered in *NEC®* Article 210. Part A covers the general provisions for branch-circuits, Part B covers the branch-circuit ratings, and Part C gives the required outlets. Outdoor branch-circuits and feeders are found in *NEC®* Article 225. *NEC®* Article 215 covers other feeders, and *NEC®* Article 220 covers the calculations for branch-circuits and feeders and the optional calculations for computing feeder and service loads. However, branch-circuits are found throughout the Code. Air-conditioning branch-circuits are found in *NEC®* Article 440, appliance branch-circuits in *NEC®* Article 424, heating branch-circuits in *NEC®* Article 422, and motor branch-circuits in *NEC®* Article 430, and so forth. Chapters 5, 6, and 7 supplement the first four chapters of the Code. Therefore, for branch-circuit studies it is necessary that you be familiar with the entire *NEC®* and use the index to quickly find the specific type branch-circuits with which you are dealing. Branch-circuits are defined in *NEC®* Article 100 and are defined as a branch-circuit, an appliance branch-circuit, a general purpose branch-circuit, an individual branch-circuit, a multiwire branch-circuit, and the branch-circuit selection current. It is important that you read these definitions before applying the requirements in the *NEC®*, so you are properly applying the requirements. A feeder is defined in Article 100 as the circuit conductor between the service equipment or the source of a separately derived system and the final branch-circuit overcurrent device.

> **Caution:** A feeder generally terminates in more than one overcurrent device.

For instance, many lighting fixtures and other equipment contain supplementary overcurrent devices at the fixture or equipment. The conductors feeding that equipment should still be considered as the branch-circuit conductors and not feeder conductors. Feeders are generally found in *NEC®* Articles 215 and 225. Calculations for sizing feeders are found in *NEC®* Article 220. Other references to feeders can be found in *NEC®* Article 364 for busways, *NEC®* Article 430

Branch-Circuit
The circuit conductors between the final over-current device protecting the circuit and the outlet(s).

Branch-Circuit, Appliance
A branch-circuit that supplied energy to one or more outlets to which appliances are to be connected, and that has no permanently connected lighting fixtures, and is not a part of an appliance.

Branch-Circuit, General-Purpose
A branch-circuit that supplies a number of outlets for lighting and appliances.

Branch-Circuit, Individual
A branch-circuit that supplies only one utilization equipment.

Branch-Circuit, Multiwire
A branch-circuit consisting of two or more ungrounded conductors having a potential difference between them, and a grounded conductor having equal potential difference between it and each ungrounded conductor of the circuit and that is connected to the neutral or grounded conductor of the system.

Reprinted with permission from NFPA 70-1999.

---

for motors, *NEC®* Article 550 for mobile homes, and *NEC®* 530 for motion picture studios. See the Index for a complete listing of where feeders can be found in the *NEC®*. (See Figure 6–1.)

General-Purpose Branch-Circuits

The simplest form of branch-circuits is general purpose branch-circuits, which supply a number of outlets for lighting and appliances, and these are in 15-, 20-, 30-, 40-, and 50-ampere sizes. (See Figure 6–2.) Where conductors or higher ampacity are used for any reason, the ampere rating or the setting of the specified overcurrent device determines the circuit classification. Multiwire branch-circuits greater than 50 amps can be permitted for non-lighting outlet loads on industrial premises where maintenance and supervision indicate that a qualified

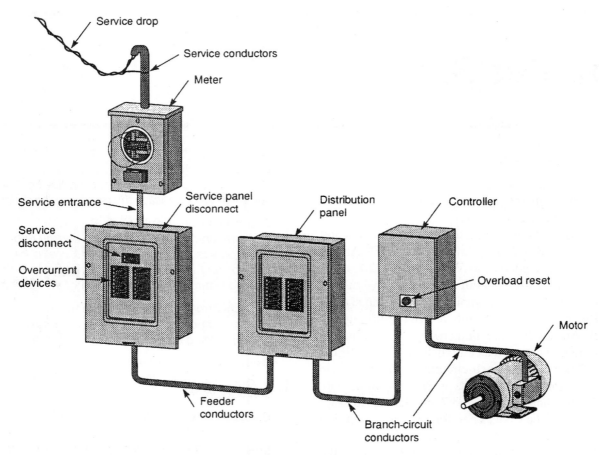

Figure 6–1   Specific conductor types as defined in *NEC*® Article 100.

person will service the equipment. A common multioutlet branch-circuit greater than 50 amperes is often found in industrial buildings using welding receptacles so the maintenance personnel can move welders throughout the plant on an as-needed basis. In these instances, you can find many receptacles on one 60-ampere or above rated circuit to supply these receptacles. Generally, no danger of overloading this circuit exists because the plant has a limited number of welding machines and personnel capable of using those machines. There are no voltage limitations on branch-circuits because branch-circuits can supply the equipment and motors of many varying different voltages.

Example: What is the maximum circuit rating for a cord- and plug-connected load that can be connected to circuit that contains two or more 15-ampere receptacles? What is the maximum load that may be connected?

Step 1. Refer to the Index "Branch-Circuits *Overcurrent Protection*," which references Section 210-20.

Step 2. Section 210-20 gives the rules for 15- and 20-ampere branch-circuits and references Table 210-21(b)(2).

Step 3. Table 210-21(b)(2).

Answer: A 15- or 20-ampere circuit with a maximum connected load of 12 amperes.

Example: How many general-use, duplex, 20-ampere receptacles can be connected to one 120-volt, 20-ampere circuit breaker?

Step 1. This question is not complete enough to reach a single answer; therefore, we must look at both residential dwelling units and at other types of installations.

Step 2. Refer to the Table of Contents: Branch-Circuits Calculations are found in Article 220.

Step 3. Computation of Branch-Circuits Section 220-3.

Step 4. Table 220-3(a). From the table we see "Dwelling units." See Section 220-3(b)(10).

Ranges. For the computation of the range loads in these examples, Column A of Table 220-19 has been used. For optional methods, see Columns B and C of Table 220-19. Except where the computations result in a major fraction of a kilowatt (0.5 or larger), such fractions are permitted to be dropped.

SI Units. For metric conversions, 1 ft^2 = 0.093 m^2 and 1 ft = 0.3048 m.

Example No. D6. Maximum Demand for Range Loads

Table 220-19, Column A applies to ranges not over 12 kW. The application of Note 1 to ranges over 12 kW (and not over 27 kW) and Note 2 to ranges over 8¾ kW (and not over 27 kW) is illustrated in the following two examples:

Ranges All the Same Rating *(see Table 220-19, Note 1)*

Assume 24 ranges, each rated 16 kW.

From Table 220-19 Column A, the maximum demand for 24 ranges of 12-kW rating is 39 kW.

16 kW exceeds 12 kW by 4.

5% × 4 = 20% (5% increase for each kW in excess of 12)

39 kW × 20% = 7.8 kW increase

39 kW + 7.8 kW = 46.8 kW (value to be used in selection of feeders)

Ranges of Unequal Rating *(see Table 220-19, Note 2)*

Assume 5 ranges, each rated 11 kW; 2 ranges, each rated 12 kW; 20 ranges, each rated 13.5 kW; 3 ranges, each rated 18 kW.

$$5 \text{ ranges} \times 12 \text{ kW} = 60 \text{ kW (use 12 kW for range rated less than 12)}$$
$$2 \text{ ranges} \times 12 \text{ kW} = 24 \text{ kW}$$
$$20 \text{ ranges} \times 13.5 \text{ kW} = 270 \text{ kW}$$
$$\underline{3 \text{ ranges} \times 18 \text{ kW} = 54 \text{ kW}}$$
$$30 \text{ ranges total kW} = \overline{408 \text{ kW}}$$

408 kW ÷ 30 ranges = 13.6 kW (average to be used for computation)

From Table 220-19 Column A, the demand for 30 ranges of 12-kW rating is 15 kW + (1 kW × 30 ranges) = 45 kW. 13.6 kW exceeds 12 kW by 1.6 kW (use 2 kW).

$$5\% \times 2 = 10\% \text{ (5\% increase for each kW in excess of 12 kW)}$$
$$45 \text{ kW} \times 10\% = 4.5 \text{ kW increase}$$
$$45 \text{ kW} + 4.5 \text{ kW} = 49.5 \text{ kW (value to be used in selection of feeders)}$$

Figure 6–2 Branch-circuit, feeder, and service load calculations for a single-family dwelling (Reprinted with permission from NFPA 70-1999, the *National Electrical Code*®, Copyright © 1998, National Fire Protection Association, Quincy, MA 02269. This reprinted material is not the complete and official position of the National Fire Protection Association on the referenced subject, which is represented only by the standard in its entirety.)

Step 5. Dwelling Occupancies. *In one-family, two-family, and multifamily dwellings and in guest rooms of hotels and motels, the outlets specified in (a), (b), and (c) are included in the general lighting load calculations of Section 220-3(a). No additional load calculations shall be required for such outlets.*

(a) All general-use receptacle outlets of 20-ampere rating or less, including receptacles connected to the circuits in Section 210-11(c)(3)

(b) The receptacle outlets specified in Sections 210-52(e) and (g)

(c) The lighting outlets specified in Sections 210-70(a) and (b)*

Answer: For dwelling units, the number is not limited.

Step 6. For all other occupancies, we must go to Section 220-3(c) because we have no special instructions under Table 220-3(b).

Step 7. Other Outlets. *Other outlets not covered in (1) through (10) shall be computed based on 180 volt-amperes per outlet.**

Step 8. I = VA ÷ E

I = 180 ÷ 120

I = 1.5 amperes

Step 9. Table 210-24 references Section 210-23(a), which limits cord- and plug-connected loads to 80% of the branch-circuit rating.

Step 10. A 20-ampere circuit breaker multiplied by 80% equals 16 amperes.

Step 11. 16 ÷ 1.5 = 10.667

Answer: For other occupancies, ten duplex receptacles are permitted.

Section 210-6 lists the limitations for branch-circuit voltages. In occupancies such as dwelling units and guest rooms of hotels, motels, and similar occupancies, the voltage is not permitted to exceed 120 volts between conductors that supply terminals of lighting fixtures and cord- and plug-connected loads 1,440 VA nominal or less, or less than ¼ horsepower, and permitted to supply terminals of medium-base, screw-shell lampholders, or lampholders of other types applied within their voltage ratings, auxiliary equipment of electric discharge lamps, cord- and plug-connected, or permanently connected utilization equipment. Circuits exceeding 120 volts nominal and not exceeding 277 volts, nominal, to ground are permitted to supply listed electric-discharge lighting fixtures equipped with medium-base screw-shell lampholder, lighting

fixtures with mogul-base screw-shell lampholders, and lampholders other than the screw-shell type applied within their voltage ratings, auxiliary equipment of electric discharge lamps, and cord- and plug-connected or permanently connected utilization equipment. Circuits exceeding 277 volts, nominal, to ground and not exceeding 600 volts, nominal, between conductors are permitted to supply auxiliary equipment of electric discharge lamps mounted in permanently installed fixtures where the fixtures are mounted in accordance with *NEC®* 210-6(d) and exceptions. (See Figure 6–3.)

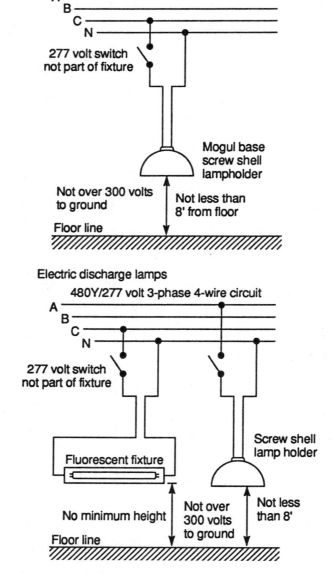

Figure 6–3 Examples of incandescent and electric discharge lighting fixtures. *(Courtesy of American Iron & Steel Institute)*

Branch-circuits supplying equipment over 600 volts are required to comply with the appropriate Code sections and articles related to that equipment.

Ground-Fault Protection

Ground-fault circuit-interrupter protection for personnel for dwelling units and other location requirements can be found in Section 210-8. However, other requirements for ground-fault circuit interrupters can be found in specific articles, such as Article 511 for commercial garages, Article 517 for health care facilities, and Article 680 for swimming pools. Section 210-52 covers the requirements and general provisions for installing receptacle outlets in dwellings and does not cover installations other than dwelling installations generally; there are some exceptions. Provisions are in the Code, in Article 220, that do cover these receptacle outlets other than dwelling installations when installed. Lighting outlets for dwellings are covered in Section 210-70. Just as for receptacle outlets, the lighting requirements are generally not specified in the *NEC®* other than for dwellings. However, the installation procedures are specified when they are installed in these installations. When calculating the loads for branch-circuit conductors, one must verify whether these loads are rated as continuous duty or as noncontinuous duty. Continuous duty is defined in Article 100. Most loads in dwellings are considered to be noncontinuous loads because they would be on for less than 3 hours at any given time. An example of a continuous load would be the lighting in a commercial or industrial establishment, which would generally be energized for more than 3 hours at a time. Most branch-circuit calculations are made in Article 220. (See Figure 6–4.) However, specific branch-circuit requirements for branch-circuits supplying specific utilization equipment are also found in the appropriate article for that equipment, such as Article 440 for air-conditioning equipment, and Article 430 for motors. (See Figure 6–5.) There also are many other specific requirements throughout the latter chapters in the *NEC®*. Several examples for calculating branch-circuits are found in the latter sections of this chapter. Article 215 covers the installation requirements and minimum size and ampacity for conductors for feeders supplying branch-circuit loads, as are computed in Article 220.

Example: Is it permissible for the required single 20-ampere laundry circuit in a dwelling to supply both an automatic washer with a nameplate of 9.8 amperes and a gas dryer with a nameplate of 7.2 amperes?

Step 1. Section 210-52(f) requires at least one receptacle outlet.

Step 2. Section 220-4(c) requires at least one 20-ampere circuit for laundry outlet(s) to supply

(Text continues on page 76)

B. Examples

Selection of Conductors. In the following examples, the results are generally expressed in amperes. To select conductor sizes, refer to the 0 to 2000 volt ampacity tables of Article 310 and the notes that pertain to such tables.

Voltage. For uniform application of Articles 210, 215, and 220, a nominal voltage of 120, 120/240, 240, and 208Y/120 volts shall be used in computing the ampere load on the conductor.

Fractions of an Ampere. Except where the computations result in a fraction of an ampere 0.5 or larger, such fractions shall be permitted to be dropped.

Power Factor. Calculations in the following examples are based, for convenience, on the assumption that all loads have the same power factor.

Ranges. For the computation of the range loads in these examples, Column A of Table 220-19 has been used. For optional methods, see Columns B and C of Table 220-19. Except where the computations result in a major fraction of a kilowatt (0.5 or larger), such fractions shall be permitted to be dropped.

SI Units: 1 sq ft = 0.093 sq m; 1 ft = 0.3048 m.

Figure 6–4 Examples of branch-circuit, feeder, and service load calculations for different types of occupancies. (Reprinted with permission from NFPA 70-1999, the *National Electrical Code®*, Copyright © 1998, National Fire Protection Association, Quincy, MA 02269. This reprinted material is not the complete and official position of the National Fire Protection Association on the referenced subject, which is represented only by the standard in its entirety.)

Example No. D1(a). One-Family Dwelling

The dwelling has a floor area of 1500 ft^2, exclusive of an unfinished cellar not adaptable for future use, unfinished attic, and open porches. Appliances are a 12-kW range and a 5.5-kW, 240-V dryer. Assume range and dryer kW ratings equivalent to kVA ratings in accordance with Sections 220-18 and 220-19.

Computed Load *[see Section 220-10]*

General Lighting Load: 1500 ft^2 at 3 VA per ft^2 = 4500 VA

Minimum Number of Branch Circuits Required
[see Section 210-11(e)(1)]

General Lighting Load: 4500 VA ÷ 120 V = 37.5 A
  This requires three 15-A, 2-wire or two 20-A, 2-wire circuits
Small Appliance Load: Two 2-wire, 20-A circuits
  [see Section 210-11(c)(1)]
Laundry Load: One 2-wire, 20-A circuit *[see Section 210-11(c)(2)]*
Bathroom Branch Circuit: One 2-wire, 20-A circuit (no additional load
  calculation is required for this circuit) *[see Section 210-11(c)(3)]*

Minimum Size Feeder Required *[see Section 220-10]*

General Lighting	4500 VA
Small Appliance	3000 VA
Laundry	1500 VA
Total	9000 VA
3000 VA at 100%	3000 VA
9000 VA − 3000 VA = 6000 VA at 35%	2100 VA
Net Load	5100 VA
Range *(see Table 220-19)*	8000 VA
Dryer *(see Table 220-18)*	5500 VA
Net Computed Load	18,600 VA

Net Computed Load for 120/240-V, 3-wire, single-phase service or feeder

18,600 VA ÷ 240 V = 77.5 A

Net computed load exceeds 10 kVA. Section 230-42(b) would require service conductors to be 100 A.

Calculation for Neutral for Feeder and Service

Lighting and Small Appliance Net Load	5100 VA
Range: 8000 VA at 70% *(see Section 220-22)*	5600 VA
Dryer: 5500 VA at 70% *(see Section 220-22)*	3850 VA
Total	14,550 VA

Computed Load for Neutral

14,550 VA ÷ 240 V = 60.6 A

Example No. D1(b). One-Family Dwelling

Assume same conditions as Example No. D1(a), plus addition of one 6-A, 230-V, room air-conditioning unit and one 12-A, 115-V, room air-conditioning unit,* one 8-A, 115-V rated waste disposer, and one 10-A, 120-V, rated dishwasher.* See Article 430 for general motors and Article 440, Part G, for air-conditioning equipment. Motors have nameplate ratings of 115 V and 230 V for use on 120-V and 240-V nominal voltage systems.
  From Example No. D1(a), feeder current is 78 A (3-wire, 240 V).
  *(For feeder neutral, use larger of the two appliances for unbalance.)

	Line A	Neutral	Line B
Amperes from Example No. D1(a)	78	61	78
One 230-V air conditioner	6	—	6
One 115-V air conditioner and 120-V			
dishwasher	12	10	10
One 115-V disposer	—	8	8
25% of largest motor	3	2	2
(see Section 430-24)			
Total amperes per line	99	81	104

Therefore, the service would be rated 110 A

Figure 6-4 *(continued)*

Example No. D2(a). Optional Calculation for One-Family Dwelling, Heating Larger than Air Conditioning [see Section 220-30]

The dwelling has a floor area of 1500 ft², exclusive of an unfinished cellar not adaptable for future use, unfinished attic, and open porches. It has a 12-kW range, a 2.5-kW water heater, a 1.2-kW dishwasher, 9 kW of electric space heating installed in five rooms, a 5-kW clothes dryer, and a 6-A, 230-V, room air-conditioning unit. Assume range, water heater, dishwasher, space heating, and clothes dryer kW ratings equivalent to kVA.

Air Conditioner kVA Calculation

$$6 \text{ A} \times 230 \text{ V} \div 1000 = 1.38 \text{ kVA}$$

This 1.38 kVA (Section 220-30(c)(1)) is less than 40% of 9 kVA of separately controlled electric heat (Section 220-3(c)(6)), so the 1.38 kVA need not be included in the service calculation.

General Load

1500 ft² at 3 VA	4500 VA
Two 20-A appliance outlet circuits at 1500 VA each	3000 VA
Laundry circuit	1500 VA
Range (at nameplate rating)	12,000 VA
Water heater	2500 VA
Dishwasher	1200 VA
Clothes dryer	5000 VA
Total	29,700 VA

Application of Demand Factor (see Section 220-30(b))

First 10 kVA of general load at 100%	10,000 VA
Remainder of general load at 40% (19.7 kVA × 0.4)	7880 VA
Total of general load	17,880 VA
9 kVA of heat at 40% (9000 VA × 0.4) =	3600 VA
Total	21,480 VA

Calculated Load for Service Size

$$21,480 \text{ VA} \div 240 \text{ V} = 89.5 \text{ A}$$

The minimum service size would be 100 A.

Feeder Neutral Load, per Section 220-22

1500 ft² at 3 VA	4500 VA
Three 20-A circuits at 1500 VA	4500 VA
Total	9000 VA
3000 VA at 100%	3000 VA
9000 VA − 3000 VA = 6000 VA at 35%	2100 VA
Subtotal	5100 VA
Range: 8 kVA at 70%	5600 VA
Clothes dryer: 5 kVA at 70%	3500 VA
Dishwasher	1200 VA
Total	15,400 VA

Calculated Load for Neutral

$$15,400 \text{ VA} \div 240 \text{ V} = 64.2 \text{ A}$$

Figure 6-4 *(continued)*

Example No. D2(b). Optional Calculation for One-Family Dwelling, Air Conditioning Larger than Heating *[see Section 220-30]*

The dwelling has a floor area of 1500 ft², exclusive of an unfinished cellar not adaptable for future use, unfinished attic, and open porches. It has two 20-A small appliance circuits, one 20-A laundry circuit, two 4-kW wall-mounted ovens, one 5.1-kW counter-mounted cooking unit, a 4.5-kW water heater, a 1.2-kW dishwasher, a 5-kW combination clothes washer and dryer, six 7-A, 230-V room air-conditioning units, and a 1.5-kW permanently installed bathroom space heater. Assume wall-mounted ovens, counter-mounted cooking unit, water heater, dishwasher, and combination clothes washer and dryer kW ratings equivalent to kVA.

Air Conditioning kVA Calculation

$$\text{Total amperes} = 6 \text{ units} \times 7 \text{ A} = 42 \text{ A}$$

$$42 \text{ A} \times 240 \text{ V} \div 1000 = 10.08 \text{ kVA (assume PF = 1.0)}$$

Load Included at 100%

Air Conditioning: Included below *(Section 220-30(c)(1))*
Space Heater: Omit *(Section 220-30(c)(5))*

General Load

1500 ft² at 3 VA	4500 VA
Two 20-A small appliance circuits at 1500 VA each	3000 VA
Laundry circuit	1500 VA
Two ovens	8000 VA
One cooking unit	5100 VA
Water heater	4500 VA
Dishwasher	1200 VA
Washer/dryer	5000 VA
Total general load	32,800 VA
First 10 kVA at 100%	10,000 VA
Remainder at 40% (22.8 kVA × 0.4 × 1000)	9120 VA
Subtotal general load	19,120 VA
Air conditioning	10,080 VA
Total	29,200 VA

Calculated Load for Service

$$29,200 \text{ VA} \div 240 \text{ V} = 122 \text{ A (service rating)}$$

Feeder Neutral Load, per Section 220-22

Assume that the two 4-kVA wall-mounted ovens are supplied by one branch circuit, the 5.1-kVA counter-mounted cooking unit by a separate circuit.

1500 ft² at 3 VA	4500 VA
Three 20-A circuits at 1500 VA	4500 VA
Subtotal	9000 VA
3000 VA at 100%	3000 VA
9000 VA − 3000 VA = 6000 VA at 35%	2100 VA
Subtotal	5100 VA

Two 4-kVA ovens plus one 5.1-kVA cooking unit = 13.1 kVA. Table 220-19 permits 55% demand factor or 13.1 kVA × 0.55 = 7200 VA feeder capacity.

Subtotal from above	5100 VA
Ovens and cooking unit: 7200 VA × 70% for neutral load	5040 VA
Clothes washer/dryer: 5 kVA × 70% for neutral load	3500 VA
Dishwasher	1200 VA
Total	14,840 VA

Calculated Load for Neutral

$$14,840 \text{ VA} \div 240 \text{ V} = 61.83 \text{ A (use 62 A)}$$

Figure 6-4   *(continued)*

Example No. D2(c). Optional Calculation for One-Family Dwelling with Heat Pump (Single-Phase, 240/120-Volt Service)
(see Section 220-30)

The dwelling has a floor area of 2000 ft², exclusive of an unfinished cellar not adaptable for future use, unfinished attic, and open porches. It has a 12-kW range, a 4.5-kW water heater, a 1.2-kW dishwasher, a 5-kW clothes dryer, and a 2½-ton (24-A) heat pump with 15 kW of backup heat.

Heat Pump kVA Calculation

$$24 \text{ A} \times 240 \text{ V} \div 1000 = 5.76 \text{ kVA}$$

This 5.76 kVA is less than 15 kVA of the backup heat; therefore, the heat pump load need not be included in the service calculation *(see Section 220-30(c))*.

General Load

2000 ft² at 3 VA	6000 VA
Two 20-A appliance outlet circuits at 1500 VA each	3000 VA
Laundry circuit	1500 VA
Range (at nameplate rating)	12,000 VA
Water heater	4500 VA
Dishwasher	1200 VA
Clothes dryer	5000 VA
Subtotal general load	33,200 VA
First 10 kVA of general load at 100%	10,000 VA
Remainder of general load at 40% (23,200 VA × 0.4)	9280 VA
Total net general load	19,280 VA

Heat Pump and Supplementary Heat*

$$240 \text{ V} \times 24 \text{ A} = 5760 \text{ VA}$$

15-kW Electric Heat:

$$5760 \text{ VA} + 15,000 \text{ VA} = 20,760 \text{ VA} = 20.76 \text{ kVA}$$
$$20.76 \text{ kVA} \times 100\% = 20.76 \text{ kVA}$$

*If supplementary heat is not on at same time as heat pump, heat pump kVA need not be added to total.

Totals

Net general load		19,280 VA
Heat pump and supplementary heat		20,760 VA
	Total	40,040 VA

Calculated Load for Service

$$40.04 \text{ kVA} \times 1000 \div 240 \text{ V} = 166.8 \text{ A}$$

This dwelling unit would be permitted to be served by a 175-A service.

Example No. D3. Store Building

A store 50 ft by 60 ft, or 3000 ft², has 30 ft of show window. There are a total of 80 duplex receptacles. The service is a 120/240 V, single phase 3-wire service. Actual connected lighting load is 8500 VA.

Computed Load *(see Section 220-10)*

Noncontinuous Loads:
Receptacle Load *(see Section 220-13)*

80 receptacles at 180 VA	14,400 VA
10,000 VA at 100%	10,000 VA
14,400 VA − 10,000 VA = 4400 VA at 50%	2,200 VA
Subtotal	12,200 VA

Figure 6-4   *(continued)*

Continuous Loads:

General Lighting*

3000 ft^2 at 3 VA per ft^2	9000 VA
Show Window Lighting	
30 ft at 200 VA per ft	6000 VA
Outside Sign Circuit [see Section 600-5(b)(3)]	1200 VA
Subtotal	16,200 VA
Subtotal from noncontinuous	12,200 VA
Total noncontinuous + continuous loads =	28,400 VA

*In the example, 125% of the actual connected lighting load (8500 VA × 1.25 = 10,625 VA) is less than 125% of the load from Table 220-3(a), so the minimum lighting load from Table 220-3(a) is used in the calculation. Had the actual lighting load been greater than the value computed from Table 220-3(a), 125% of the actual connected lighting load would have been used.

Minimum Number of Branch Circuits Required

General Lighting: Branch circuits need only be installed to supply the actual connected load [see Section 210-11(d)].

$$8500 \text{ VA} \times 1.25 = 10,625 \text{ VA}$$

$$10,625 \text{ VA} \div 240 \text{ V} = 44 \text{ A for 3-wire, 120/240 V}$$

The lighting load would be permitted to be served by 2-wire or 3-wire, 15- or 20-A circuits with combined capacity equal to 44 A or greater for 3-wire circuits or 88 A or greater for 2-wire circuits. The feeder capacity as well as the number of branch-circuit positions available for lighting circuits in the panelboard must reflect the full calculated load of 9000 VA × 1.25 = 11,250 VA.

Show Window

$$6000 \text{ VA} \times 1.25 = 7500 \text{ VA}$$

$$7500 \text{ VA} \div 240 \text{ V} = 31 \text{ A for 3-wire, 120/240 V}$$

The show window lighting is permitted to be served by 2-wire or 3-wire circuits with a capacity equal to 31 A or greater for 3-wire circuits or 62 A or greater for 2-wire circuits.

Receptacles required by Section 210-62 are assumed to be included in the receptacle load above if these receptacles do not supply the show window lighting load.

Receptacles

Receptacle Load: 14,400 VA ÷ 240 V = 60 A for 3-wire, 120/240 V

The receptacle load would be permitted to be served by 2-wire or 3-wire circuits with a capacity equal to 60 A or greater for 3-wire circuits or 120 A or greater for 2-wire circuits.

Minimum Size Feeder (or Service) Overcurrent Protection
[see Sections 215-3 or 230-90(a)]

Subtotal noncontinuous loads		12,200 VA
Subtotal continuous load at 125%		
(16,200 VA × 1.25)		20,250 VA
	Total	32,450 VA

$$32,450 \text{ VA} \div 240 \text{ V} = 135 \text{ A}$$

The next higher standard size is 150 A (see Section 240-6).

Minimum Size Feeders (or Service Conductors) Required
[see Sections 215-2, 215-3, and 230-42(a)]

For 120/240-V, 3-wire system,

$$32,450 \text{ VA} \div 240 \text{ V} = 135 \text{ A}$$

Service or feeder conductor is 1/0 Cu per Section 215-3 and Table 310-16 (with 75°C terminations).

Figure 6-4 *(continued)*

Example No. 4(a). Multifamily Dwelling

Multifamily dwelling having 40 dwelling units.

Meters in two banks of 20 each and individual feeders to each dwelling unit.

One-half of the dwelling units are equipped with electric ranges not exceeding 12 kW each. Assume range kW rating equivalent to kVA rating in accordance with Section 220-19. Other half of ranges are gas ranges.

Area of each dwelling unit is 840 sq ft.

Laundry facilities on premises available to all tenants. Add no circuit to individual dwelling unit. Add 1500 volt-amperes for each laundry circuit to house load and add to the example as a "house load."

Computed Load for Each Dwelling Unit (Article 220):

General Lighting Load:

840 sq ft at 3 volt-amperes per sq ft	2520 volt-amperes

Special Appliance Load:

Electric Range (Section 220-19)	8000 volt-amperes

Minimum Number of Branch Circuits Required for Each Dwelling Unit
(Section 220-4):

General Lighting Load: 2520 volt-amperes ÷ 120 volts = 21 amperes or two 15-ampere, 2-wire circuits; or two 20-ampere, 2-wire circuits.

Small Appliance Load: Two 2-wire circuits of No. 12 wire. [See Section 220-4(b).]

Range Circuit: 8000 volt-amperes ÷ 240 volts = 33 amperes or a circuit of two No. 8 conductors and one No. 10 conductor as permitted by Section 220-22. (See Section 210-19.)

Minimum Size Feeder Required for Each Dwelling Unit
(Section 215-2):

Computed Load (Article 220):

General Lighting Load	2520 volt-amperes
Small Appliance Load: two 20-ampere circuits	3000 volt-amperes
Total Computed Load (without ranges)	5520 volt-amperes

Application of Demand Factor:

3000 volt-amperes at 100%	3000 volt-amperes
2520 volt-amperes at 35%	882 volt-amperes
Net Computed Load (without ranges)	3882 volt-amperes
Range Load	8000 volt-amperes
Net Computed Load (with ranges)	11,882 volt-amperes

Size of Each Feeder (see Section 215-3).

For 120/240-volt, 3-wire system (without ranges):

Net Computed Load, 3882 volt-amperes ÷ 240 volts = 16.2 amperes.

For 120/240-volt, 3-wire system (with ranges):

Net Computed Load, 11,882 volt-amperes ÷ 240 volts = 49.5 amperes.

Feeder Neutral:

Lighting and Small Appliance Load	3882 volt-amperes
Range Load: 8000 volt-amperes at 70% (see Section 220-22)	5600 volt-amperes
(Not included for apartments without electric range)	
Net Computed Load (neutral)	9482 volt-amperes

9482 volt-amperes ÷ 240 volts = 39.5 amperes

Minimum Size Feeders Required from Service Equipment to Meter Bank (For 20 Dwelling Units — 10 with Ranges):

Total Computed Load:

Lighting and Small Appliance Load:

20 units × 5520 volt-amperes	110,400 volt-amperes

Application of Demand Factor:

3000 volt-amperes at 100%	3000 volt-amperes
107,400 volt-amperes at 35%	37,590 volt-amperes
Net Computed Lighting and Small Appliance Load	40,590 volt-amperes
Range Load, 10 ranges (less than 12 kVA, Col. A, Table 220-19)	25,000 volt-amperes
Net Computed Load (with ranges)	65,590 volt-amperes

For 120/240-volt, 3-wire system:

Net Computed Load, 65,590 volt-amperes ÷ 240 volts = 273 amperes

Feeder Neutral:

Lighting and Small Appliance Load	40,590 volt-amperes
Range Load: 25,000 volt-amperes at 70% (see Section 220-22)	17,500 volt-amperes
Computed Load (neutral)	58,090 volt-amperes

58,090 volt-amperes ÷ 240 volts = 242 amperes

Figure 6-4   *(continued)*

Example No. 4(a) (continued)

Further Demand Factor (Section 220-22):

200 amperes at 100%	200 amperes
42 amperes at 70%	29 amperes
Net Computed Load (neutral)	229 amperes

Minimum Size Main Feeder (or Service Conductors) Required (less house load) (For 40 Dwelling Units—20 with Ranges):

Total Computed Load:

Lighting and Small Appliance Load:	
40 units × 5520 volt-amperes	220,800 volt-amperes
Application of Demand Factor:	
3000 volt-amperes at 100%	3000 volt-amperes
117,000 volt-amperes at 35%	40,950 volt-amperes
100,800 volt-amperes at 25%	25,200 volt-amperes
Net Computed Lighting and Small Appliance Load	69,150 volt-amperes
Range Load, 20 ranges (less than 12 kVA, Col. A, Table 220-19)	35,000 volt-amperes
Net Computed Load	104,150 volt-amperes

For 120/240-volt, 3-wire system:

Net Computed Load, 104,150 volt-amperes ÷ 240 volts = 434 amperes

Feeder Neutral:

Lighting and Small Appliance Load	69,150 volt-amperes
Range Load: 35,000 volt-amperes at 70% (see Section 220-22)	24,500 volt-amperes
Computed Load (neutral)	93,650 volt-amperes

93,650 volt-amperes ÷ 240 volts = 390 amperes.

Further Demand Factor (see Section 220-22):

200 amperes at 100%	200 amperes
190 amperes at 70%	133 amperes
Net Computed Load (neutral)	333 amperes

See Tables 310-16 through 310-19, Notes 8 and 10.

Example No. 4(b). Optional Calculation for Multifamily Dwelling

Multifamily dwelling equipped with electric cooking and space heating or air conditioning and having 40 dwelling units.

Meters in two banks of 20 each plus house metering and individual feeders to each dwelling unit.

Each dwelling unit is equipped with an electric range of 8-kW nameplate rating, four 1.5-kW separately controlled 240-volt electric space heaters, and a 2.5-kW 240-volt electric water heater. Assume range, space heater, and water heater kW ratings equivalent to kVA.

A common laundry facility is available to all tenants [Section 210-52(f), Exception No. 1].

Area of each dwelling unit is 840 sq ft.

Computed Load for Each Dwelling Unit (Article 220):

General Lighting Load:

840 sq ft at 3 volt-amperes per sq ft	2520 volt-amperes
Electric Range	8000 volt-amperes
Electric Heat 6 kVA (or air conditioning if larger)	6000 volt-amperes
Electric Water Heater	2500 volt-amperes

Minimum Number of Branch Circuits Required for Each Dwelling Unit:

General Lighting Load: 2520 volt-amperes ÷ 120 volts = 21 amperes or two 15-ampere, 2-wire circuits, or two 20-ampere, 2-wire circuits.

Small Appliance Loads: Two 2-wire circuits of No. 12 [see Section 220-4(b)].

Range Circuit: 8000 volt-amperes × 80% ÷ 240 volts = 27 amperes on a circuit of three No. 10 conductors as permitted in Column C of Table 220-19.

Space Heating 6000 volt-amperes ÷ 240 volts = 25 amperes

No. of circuits (see Section 220-4).

Minimum Size Feeder Required for Each Dwelling Unit (Section 215-2):

Computed Load (Article 220):

General Lighting Load	2520 volt-amperes
Small Appliance Load, two 20-ampere circuits	3000 volt-amperes
Total Computed Load (without range and space heating) ..	5520 volt-amperes
Application of Demand Factor:	
3000 volt-amperes at 100%	3000 volt-amperes
2520 volt-amperes at 35%	882 volt-amperes
Net Computed Load (without range and space heating) ...	3882 volt-amperes

Figure 6-4 *(continued)*

Example No. 4(b) (continued)

Range Load	6400 volt-amperes
Space Heating (Section 220-15)	6000 volt-amperes
Water Heater	2500 volt-amperes
Net Computed Load for individual dwelling unit	18,782 volt-amperes

For 120/240-volt, 3-wire system

Net Computed Load 18,782 volt-amperes ÷ 240 volts = 78 amperes

Feeder Neutral (Section 220-22)

Lighting and Small Appliance Load	3882 volt-amperes
Range Load: 6400 volt-amperes at 70% (see Section 220-22)	4480 volt-amperes
Space and Water Heating (no neutral) 240 volts	0 volt-amperes
Net Computed Load (neutral)	8362 volt-amperes

8362 volt-amperes ÷ 240 volts = 35 amperes

Minimum Size Feeder Required from Service Equipment to Meter Bank for 20 Dwelling Units:

Total Computed Load:

Lighting and Small Appliance Load 20 units × 5520 volt-amperes	110,400 volt-amperes
Water and Space Heating Load 20 units × 8500 volt-amperes	170,000 volt-amperes
Range Load 20 × 8000 volt-amperes	160,000 volt-amperes
Net Computed Load (20 dwelling units)	440,400 volt-amperes
Net Computed Load Using Optional Calculation (Table 220-32)	
440,400 volt-amperes × .38	167,352 volt-amperes

167,352 volt-amperes ÷ 240 volts = 697 amperes

Minimum Size Main Feeder Required (less house load) for 40 Dwelling Units:

Total Computed Load:

Lighting and Small Appliance Load 40 units × 5520 volt-amperes	220,800 volt-amperes
Water and Space Heating Load 40 units × 8500 volt-amperes	340,000 volt-amperes
Range Load 40 × 8000 volt-amperes	320,000 volt-amperes
Net Computed Load (40 dwelling units)	880,800 volt-amperes
Net Computed Load Using Optional Calculation (Table 220-32)	
880,800 volt-amperes × 0.28	246,624 volt-amperes

246,624 volt-amperes ÷ 240 volts = 1028 amperes

Feeder Neutral Load for Feeder from Service Equipment to Meter Bank for 20 Dwelling Units:

Lighting and Small Appliance Load

20 units × 5520 volt-amperes	110,400 volt-amperes
First 3000 volt-amperes at 100%	3000 volt-amperes
107,400 volt-amperes at 35%	37,590 volt-amperes
Subtotal	40,590 volt-amperes
20 Ranges = 35,000 volt-amperes at 70% (See Table 220-19 and Section 220-22.)	24,500 volt-amperes
Total	65,090 volt-amperes

65,090 volt-amperes ÷ 240 volts = 271 amperes

Further Demand Factor (Section 220-22)

First 200 amperes at 100%	200 amperes
Balance: 71 amperes at 70%	50 amperes
Total	250 amperes

Feeder Neutral Load of Main Feeder (less house load) for 40 Dwelling Units:

Lighting and Small Appliance Load

40 units × 5520 volt-amperes	220,800 volt-amperes
First 3000 volt-amperes at 100%	3000 volt-amperes
120,000 volt-amperes - 3000 volt-amperes = 117,000 volt-amperes at 35%	40,950 volt-amperes
220,800 volt-amperes - 120,00 volt-amperes = 100,800 volt-amperes at 25%	25,200 volt-amperes
Net Computed Lighting and Small Appliance Load	69,150 volt-amperes
40 Ranges = 55,000 volt-amperes at 70% (See Table 220-19 and Section 220-22)	38,500 volt-amperes
Total	107,650 volt-amperes

107,650 volt-amperes ÷ 240 volts = 449 amperes

Further Demand Factor (Section 220-22)

First 200 amperes at 100%	200 amperes
Balance: 249 amperes at 70%	174 amperes
Total	374 amperes

Figure 6-4 *(concluded)*

Example No. D8. Motor Circuit Conductors, Overload Protection, and Short-Circuit and Ground-Fault Protection
(see Sections 240-6, 430-6, 430-22, 430-23, 430-24, 430-32, 430-34, 430-52, and 430-62, Tables 430-150 and 430-152)

Determine the minimum required conductor ampacity, the motor overload protection, the branch-circuit short-circuit and ground-fault protection, and the feeder protection, for three induction-type motors on a 480-V, 3-phase feeder, as follows:

(a) One 25-hp, 460-V, 3-phase, squirrel-cage motor, nameplate full-load current 32 A, Design B, Service Factor 1.15

(b) Two 30-hp, 460-V, 3-phase, wound-rotor motors, nameplate primary full-load current 38 A, nameplate secondary full-load current 65 A, 40°C rise.

Conductor Ampacity

The full-load current value used to determine the minimum required conductor ampacity is obtained from Table 430-150 *[see Section 430-6(a)]* for the squirrel-cage motor and the primary of the wound-rotor motors. To obtain the minimum required conductor ampacity, the full-load current is multiplied by 1.25 *[see Sections 430-22 and 430-23(a)]*.

For the 25-hp motor,

$$32 \text{ A} \times 1.25 = 40 \text{ A}$$

For the 30-hp motors,

$$38 \text{ A} \times 1.25 = 47.5 \text{ A}$$
$$65 \text{ A} \times 1.25 = 81.25 \text{ A}$$

Motor Overload Protection

Where protected by a separate overload device, the motors are required to have overload protection rated or set to trip at not more than 125% of the nameplate full-load current *[see Sections 430-6(a) and 430-32(a)(1)]*.

For the 25-hp motor,

$$32 \text{ A} \times 1.25 = 40.0 \text{ A}$$

For the 30-hp motors,

$$38 \text{ A} \times 1.25 = 47.5 \text{ A}$$

Where the separate overload device is an overload relay (not a fuse or circuit breaker), and the overload device selected at 125% is not sufficient to start the motor or carry the load, the trip setting is permitted to be increased in accordance with Section 430-34.

Branch-Circuit Short-Circuit and Ground-Fault Protection

The selection of the rating of the protective device depends on the type of protective device selected, in accordance with Section 430-52 and Table 430-152. The following is for the 25-hp motor.

Nontime-Delay Fuse: The fuse rating is 300% × 32 A = 96 A.
    The next larger standard fuse is 100 A *[see Sections 240-6 and 430-52(c)(1), Exception No. 1]*. If the motor will not start with a 100-A nontime-delay fuse, the fuse rating is permitted to be increased to 125 A because this rating does not exceed 400% *[see Section 430-52(c)(1), Exception No. 2a]*.

Time-Delay Fuse: The fuse rating is 175% × 32 A = 56 A.
    The next larger standard fuse is 60 A *[see Sections 240-6 and 430-52(c)(1), Exception No. 1]*. If the motor will not start with a 60-A time-delay fuse, the fuse rating is permitted to be increased to 70 A because this rating does not exceed 225% *[see Section 430-52(c)(1), Exception No. 2b]*.

Feeder Short-Circuit and Ground-Fault Protection

The rating of the feeder protective device is based on the sum of the largest branch-circuit protective device (example is 100 A) plus the sum of the full-load currents of the other motors, or 100 A + 40 A + 40 A = 180 A. The nearest standard fuse that does not exceed this value is 175 A *[see Sections 240-6 and 430-62(a)]*.

Figure 6–5   Examples of motor calculations. (Reprinted with permission from NFPA 70-1999, the *National Electrical Code®*, Copyright © 1998, National Fire Protection Association, Quincy, MA 02269. This reprinted material is not the complete and official position of the National Fire Protection Association on the referenced subject, which is represented only by the standard in its entirety.)

Example No. 9 (continued)

Conductor ampacity is determined as follows:

(a) Per Sections 620-13(d) and 620-61(b)(1), use Table 430-22(a), Exception, for intermittent duty (elevators). For intermittent duty using a continuous rated motor, the percentage of nameplate current rating to be used is 140 percent.

(b) For the 30-horsepower ac drive motor, 1.4 × 40 amperes = 56 amperes. For the 40-horsepower ac drive motor, 1.4 × 52 amperes = 73 amperes.

(c) The total conductor ampacity is the sum of all the motor currents — (73 amperes) + (5 × 56 amperes) = 353 amperes.

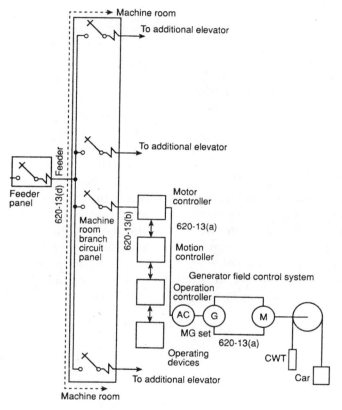

Example No. 9 Figure

(d) Per Section 620-14 and Table 620-14, the conductor (feeder) ampacity shall be permitted to be reduced by the use of a demand factor. Constant loads are not included. (See Section 620-14, FPN.) For six elevators, the demand factor is 0.79. The feeder diverse ampacity is, therefore, 0.79 × 353 amperes = 279 amperes.

(e) Per Sections 430-24 and 220-10(b), the controller continuous current is 1.25 × 10 amperes = 12.5 amperes.

(f) The total feeder ampacity is the sum of the diverse current and all the controller continuous current. I total = 279 amperes + (6 × 12.5 amperes) = 354 amperes.

(g) This ampacity shall be permitted to be used to select the wire size. See Diagram for Example No. 9.

Figure 6-5   *(concluded)*

the receptacle(s) required by 210-52(f) Section 422-4(b). Also, see references 210-23.

> **NOTE:** This means all outlets for laundry purposes, i.e., washer, 120-volt dryer, and ironing closet.

Step 3. Section 210-24 and Table 210-24 references Section 210-23(a).

Step 4. Section 210-23(a). The rating of any one utilization equipment supplied from a cord and plug shall not exceed 80% of the branch-circuit rating.

Step 5. 20 × 80% = 16 amperes. Therefore, that requirement is acceptable.

Step 6. 9.8 + 7.2 = 17 amperes. The rating of the circuit is not exceeded.

Answer: Yes. Reference 422-4(b), 210-24, 210-23, 210-23(a), and 210-4(b). Also, see definition of *continuous load*. These would not be considered a continuous load.

Outside Feeders and Branch-Circuits

Article 225 covers outside branch-circuits and feeders. Article 225 covers the electrical equipment and wire for the supply of utilization equipment located on or attached to the outside of public and private buildings, or run between buildings, other structures or poles on premises served. For clearances, see Figure 6–6. An example of a feeder would be those conductors being fed from an overcurrent device in the service equipment to a subpanel in another part of the building or feeding a separate building on the premises, terminating in a panel or group of overcurrent devices within the second building or in another location. A feeder also serves separately derived systems in many cases, such as a transformer where the voltage is reduced to serve lighting and receptacle branch-circuit loads. The 1999 *NEC*® relocated Section 225-8 to Section 225-32. This section provides that where more than one building or structure is on the same property and under single management, each building or other structure served shall be provided with a means of disconnecting all ungrounded conductors, which must be installed either inside or outside the building or structure in a readily accessible location nearest the point of entrance of the supply feeder conductors. The disconnect disconnecting these feeder conductors must be installed in accordance with Sections 230-70 and 230-72 and must be suitable as service equipment. However, they are required to have overcurrent protection in accordance with Article 220 for branch-circuits and Article 440 for feeders. The reference to Article 220 is necessary because residential buildings and structures such as garages and other outbuildings are permitted to be supplied with branch-circuits where the loads permit. Examples for calculating feeders and outside branch-circuits are shown in this chapter.

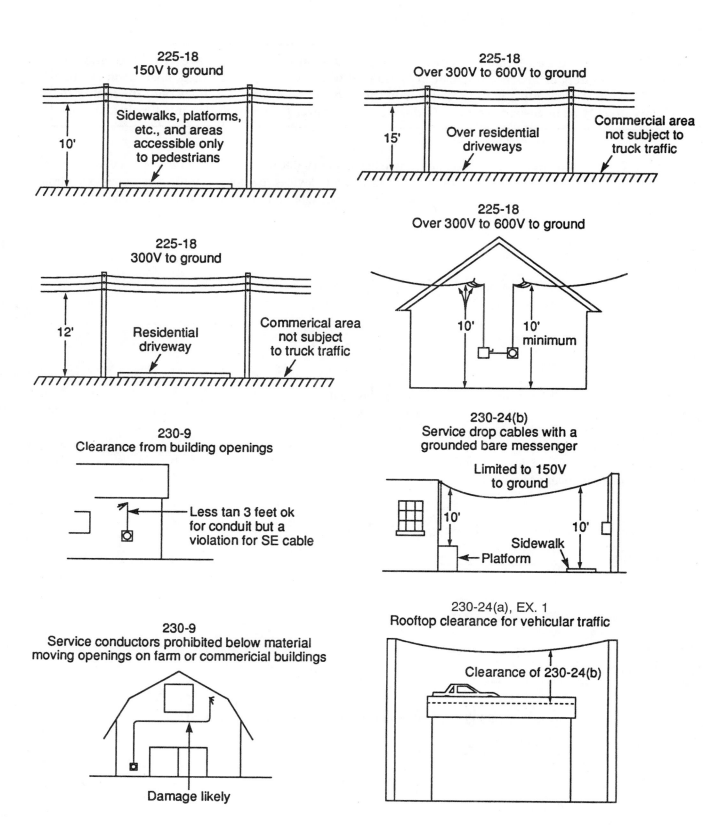

Figure 6-6 Required clearances for service and feeder conductors

CHAPTER 6 QUESTION REVIEW

> **Each lesson is designed purposely to require the student to apply the entire *NEC*® text, and not specific chapters or articles. It has been found that when studying for a timed, open-book examination, the student must gain proficiency in the Table of Contents, the Index, and the ability to move quickly from cover to cover to find the correct answer to each question in a timely fashion.**

1. Fluorescent lighting fixtures, each containing two ballasts rated 0.8 amperes each at 120 volts, are to be installed for general lighting in a store. The overcurrent protection devices are not listed for continuous operation at 100% of its rating. What is the maximum number of these lighting fixtures that may be permanently wired to a 20 amp, 120-volt branch-circuit?

   Answer: _____

   _____

   _____

   _____

   Reference: _____

Example for Questions 2 through 20: You are going to build an investment single-family dwelling 28 feet by 40 feet. There will be three bedrooms and one bathroom with central natural gas heat located under the floor in the crawlspace. There will be no air conditioning. There will be an electric water heater (40-gallon, quick recovery) with two 4,500-watt elements, a gas cooking range, and electric dryer. There will be a detached single-car garage. The electrical installation will be made in accordance with the 1999 *NEC*®. The attic in this dwelling is not suitable for the location of equipment or for storage.

2. How many square feet will this house contain?

   Answer: _____

   _____

   _____

   _____

   Reference: _____

3. How many 15-ampere general lighting and receptacle branch-circuits are required?

   Answer: _____

   _____

   _____

   _____

   Reference: _____

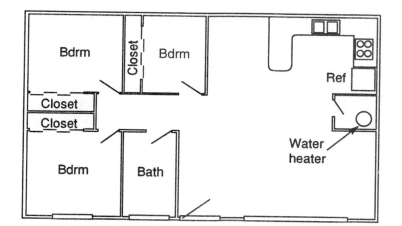

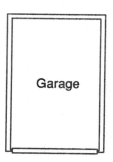

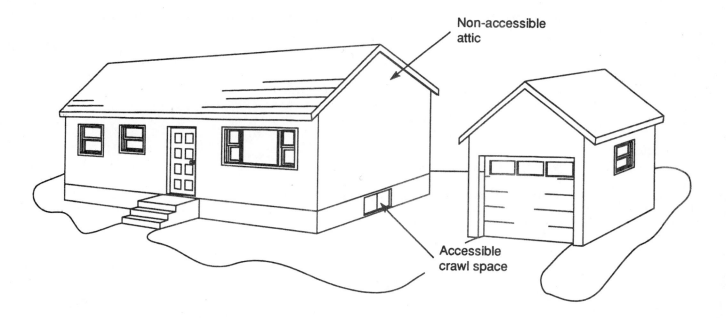

4. How many 20-ampere lighting and receptacle branch-circuits are required?

Answer: _____

_____

_____

_____

Reference: _____

5. How many appliance branch-circuits are required?

    Answer: _____

    _____

    _____

    _____

    Reference: _____

6. Is a special circuit required for the laundry?

    Answer: _____

    _____

    _____

    _____

    Reference: _____

7. What is the maximum standard size overcurrent device permitted to protect the electric water heater?

    Answer: _____

    _____

    _____

    _____

    Reference: _____

8. How many branch-circuits are required to supply the detached garage?

    Answer: _____

    _____

    _____

    _____

    Reference: _____

9. How many outdoor receptacles are required?

    Answer: _____

    _____

    _____

    _____

    Reference: _____

10. What is the minimum number of ground-fault circuit-interrupter (GFCI) devices required?

   Answer: _____

   _____

   _____

   _____

   Reference: _____

11. Are lighting outlets required at each outside door?

   Answer: _____

   _____

   _____

   _____

   Reference: _____

12. Are lighting outlets required in the attic not used for storage?

   Answer: _____

   _____

   _____

   _____

   Reference: _____

13. Is a receptacle required under the floor containing equipment requiring servicing?

   Answer: _____

   _____

   _____

   _____

   Reference: _____

14. If the answer to Question 6 above is yes, what size branch-circuit would be required?

   Answer: _____

   _____

   _____

   _____

   Reference: _____

15. Would it be permissible to connect the furnace to a general lighting and receptacle branch-circuit provided it does not overload the circuit?

    Answer: _____

    _____

    _____

    _____

    Reference:_____

16. Are lighting outlets required in each closet?

    Answer: _____

    _____

    _____

    _____

    Reference:_____

17. Is it permissible to connect the electric ignition system for gas ranges to one of the required small-appliance branch-circuits?

    Answer: _____

    _____

    _____

    _____

    Reference:_____

18. If an outdoor receptacle is located more than 6 feet above the ground level, is it required to be GFCI protected?

    Answer: _____

    _____

    _____

    _____

    Reference:_____

19. Could the outdoor receptacle described in Question 18 serve as the required outdoor receptacle?

    Answer: _____

    _____

    _____

    _____

    Reference:_____

20. What is the minimum size service required for this dwelling? (Calculate answers where required.)
    (a) General-purpose lighting and receptacle load
    (b) Appliance branch-circuit load
    (c) Water heater
    (d) Laundry
    (e) Dryer
    (f) Detached garage
    (g) Furnace
    (h) Outdoor receptacle
    (i) Air conditioning
    (j) Range
    (k) Bathroom
    (l) Other

(a) _____

(b) _____

(c) _____

(d) _____

(e) _____

(f) _____

(g) _____

(h) _____

(i) _____

(j) _____

(k) _____

(l) _____

Chapter Seven

SERVICES 600 VOLTS OR LESS

NEC® Article 230 covers the requirements for services, service conductors, and equipment for the control and protection of services and their installation requirements. Parts A through G, cover services, 600 volts, nominal, or less, Part H has additional requirements for services exceeding 600 volts, nominal. It should be remembered that all of Article 230 applies to these services, and Part H is additional provisions that supplement or modify the rest of Article 230. Part H shall not apply to the equipment on the supply side of the service point. Clearances for conductors over 600 volts are found in ANSI C2, the *National Electrical Safety Code.* Section 230-2 generally limits a building or structure to one service. There are seven exceptions that permit more than one service per building or structure. It is necessary to comply with one of these exceptions any time two or more services are being installed on a building or structure. Confusing to many are the per-

missiveness for more than one set of service conductors permitted by Exceptions 1 and 2 of Section 230-40. These exceptions frequently are used on strip shopping centers, condominium complexes, and apartment buildings. The rule satisfies the needs for individual supply, control, and metering of occupancies in these multi-occupancy buildings. However, it does not relieve the requirement of 230-2 for one service for each building or structure. (See Figure 7–1 and Figure 7–2.) The service conductors, as defined in Article 100, are required to go directly to the disconnecting means of that building or structure, because the disconnecting means is required to be installed at a readily accessible location either outside of the building or structure, or inside nearest the point of the entrance of the service conductors. Section 230-6 defines that conductors shall be considered outside of a building or other structure where installed under not less than 2 inches of concrete beneath a building or structure, where installed within a building or structure in a raceway that is encased in

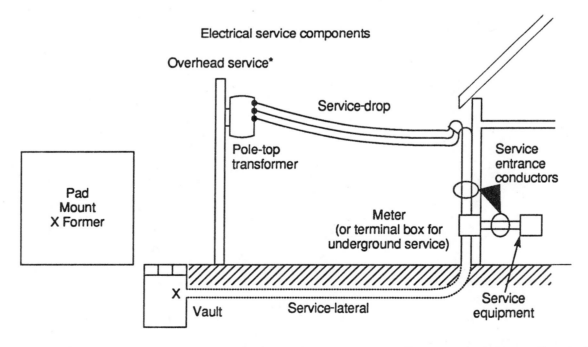

Figure 7–1 *NEC®* Section 230-2 permits only one service to each building. There are seven exceptions to this rule. The service may be supplied with aerial spans overhead or by underground "service-laterals" from the "service point." (See Definition Article 100.) **Note:** A vault is not required for underground runs but may be installed where necessary.

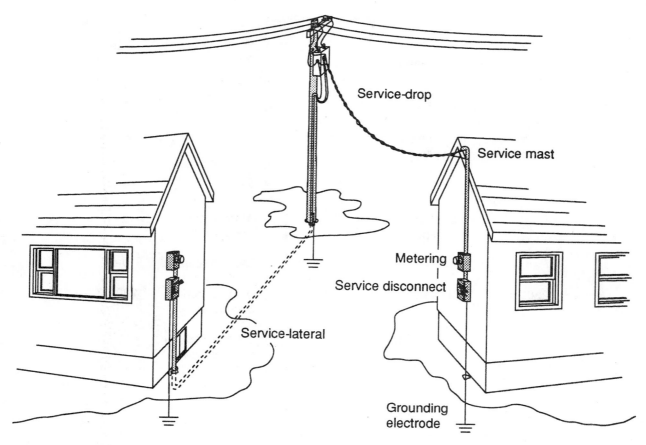

Figure 7–2 Illustration of two common types of residential services, overhead service-drop and underground service-lateral

concrete or brick not less than 2 inches thick, or where installed in a transformer vault that complies with the requirements of *NEC®* Article 450, Part C. A typical service supplying a building or structure contains several carefully defined parts. These parts are defined in Article 100 as service, service cable, service conductors, service-drop, service-entrance conductors, overhead system, underground system, service-lateral, service equipment, and service point. Each service will contain several of these parts, however, not all of these parts. It is important that the components being installed in accordance with the *NEC®* be correctly defined before sizing, calculating, or installing. For instance, a typical residential service can be supplied by the serving utility either overhead or underground. When supplied overhead, it shall be supplied as a service-drop. When supplied underground, it shall be supplied as a service-lateral. These terms should be studied carefully in *NEC®* Article 100 so calculations and design can be made correctly. The diagram in Section 230-1 should help the reader identify and use these terms correctly. (See Figure 7–3.)

Part B of Article 230 covers the overhead service-drop conductors, size and rating, and clearances required for making this installation. These conductors are generally installed by the serving utility; however, it is the responsibility of the installer to locate the service mast so these clearances will be met. Careful and close coordination between the serving utility, the customer to be served, and the installer are necessary to ensure compliance of this section of the *NEC®*. (See Figure 6–6 for clearance examples.)

Part C of Article 230 covers underground service-laterals. (See Figure 7–4.) Care should be taken so that the service-lateral is installed correctly and so that where it emerges from the ground it is amply protected by rigid or IMC steel conduit or Schedule 80 PVC to provide these conductors with physical protection against damage. Article 230, Part D covers the requirements for the service-entrance conductors. These service-entrance conductors, as defined in *NEC®* Article 100, must be installed in one of the wiring methods listed in Section 230-43, and sized and rated in accordance with Section 230-42. Specific requirements for

Article 230 – Services

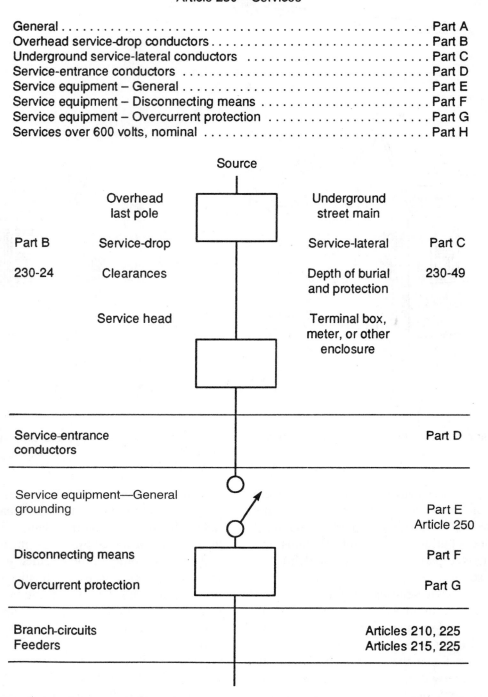

Figure 7–3 Article 230 is divided into several parts. Each covers a specific part of service installation. (Reprinted with permission from NFPA 70-1999, the *National Electrical Code®*, Copyright © 1998, National Fire Protection Association, Quincy, MA 02269. This reprinted material is not the complete and official position of the National Fire Protection Association on the referenced subject, which is represented only by the standard in its entirety.)

these service-entrance conductors are covered in Article 230 Part D and must be complied with regardless of how they are routed. The general requirements for service equipment can be found in Article 230, Part E, and the disconnecting means is found in Part F.

Each service disconnecting means is required to be suitable for the prevailing equipment conditions, and when installed in hazardous locations must comply with Chapter 5 of the *National Electrical Code®*. Each service disconnecting means permitted by Section 230-

Electrical service components

Overhead service*

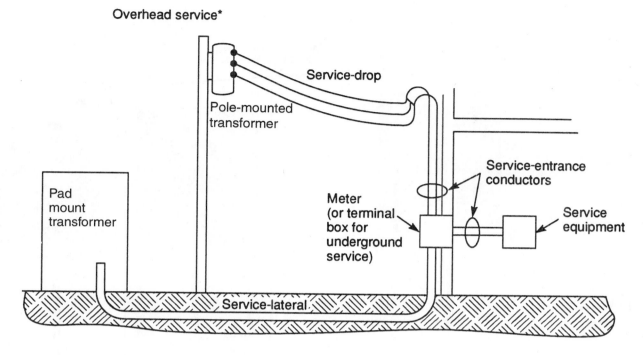

Figure 7–4 *NEC®* Article 230-2 permits only one service per building. It is permissible to supply overhead or underground.

2 or for each set of service-entrance conductors permitted by Section 230-40, Exception 1 (see Figure 7–5) shall consist of not more than six switches or six circuit breakers mounted in a single enclosure, in a group of separate enclosures, or on a switchboard. (See Figure 7–6.) There shall be no more than six disconnects per service grouped in any one location. These requirements are critical and must be followed. The disconnecting means must be grouped; each disconnect must be marked to indicate the load served. An exception permits one of the six disconnecting means permitted where used for a water pump intended to provide fire protection to be located remotely from the other disconnecting means. In multiple-occupancy buildings, each occupant shall have access unless local building management, which supplies continuous supervision, is present. In such a case, the service disconnecting means shall be permitted to be accessible to the authorized management personnel only. Each service disconnecting means shall simultaneously disconnect the ungrounded service conductors from the premise wiring. Where the service disconnecting means does not disconnect the grounded conductor from the premise wiring, other means shall be provided for this purpose in the service equipment. A terminal or bus to which all grounded conductors can be attached by means of pressure connections shall be permitted for this purpose.

Example: A 277/480-volt, 3-phase service to a building is rated at 1,600 amperes. The required main disconnecting means consists of two 800-ampere fused disconnects. Is this service required to have ground-fault protection (GFP)?

Step 1. Index "Ground-Fault Protection." Ground-fault protection for service disconnects Section 230-95.

NOTE: To answer this question, some basic electrical knowledge is required. First, the voltage tells the story. The number 277/480 indicates it is first a solidly grounded system. It also indicates it is a wye system because 480 equals 1.73% of 277. This information can be obtained from additional reference books listed as study material.

Answer: No, reference Sections 230-71. The maximum number of service disconnects is six (6). Section 230-95 states that solidly grounded wye services of more than 150 volts to ground not exceeding 600 volts phase-to-phase and over 1,000 amperes, requires ground-fault protection.

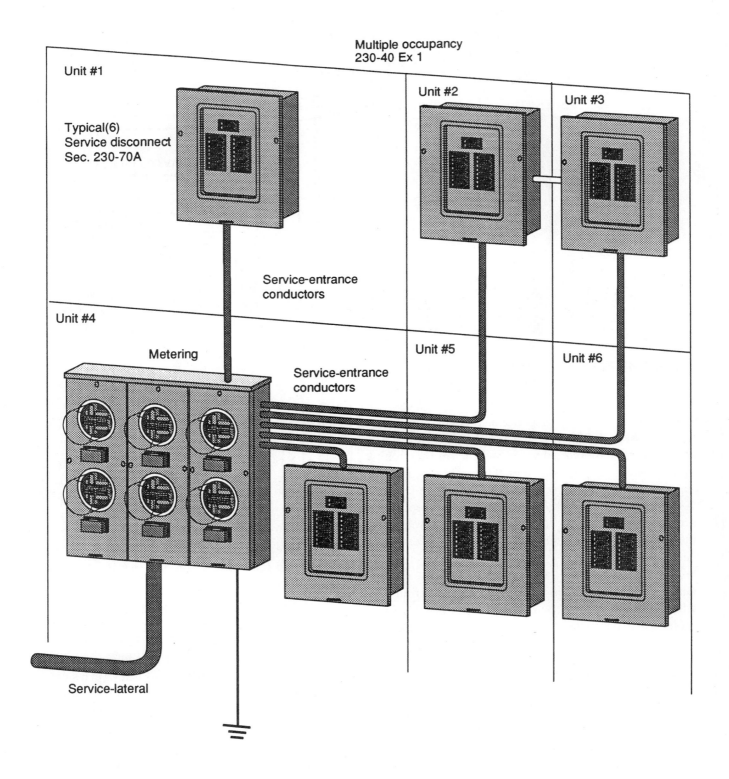

Multiple occupancy
230-40 Ex 1

Unit #1

Typical(6)
Service disconnect
Sec. 230-70A

Unit #2

Unit #3

Service-entrance
conductors

Unit #4

Metering

Service-entrance
conductors

Unit #5

Unit #6

Service-lateral

Figure 7–5 Typical riser sketch of a multi-occupancy building. Section 230-40 Exception 1 permits the main disconnecting means for that occupancy to be located on the load end of the feeder within the occupancy.

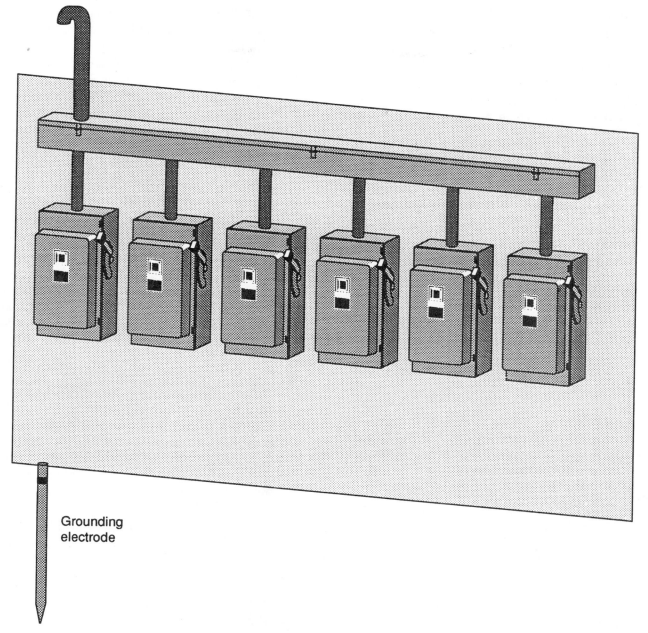

Grounding electrode

Figure 7–6 Typical sketch of six individual disconnecting means on one service as permitted by Section 230-71(a).

Part G covers the overcurrent protection and the location of the overcurrent protection requirements for services. All solidly grounded wye services of more than 150 volts to ground but not exceeding 600 volts phase-to-phase rated 1,000 amps or more must be provided with ground-fault protection of equipment. Two exceptions to this are: Exception 1 states ground-fault protection does not apply to the service disconnecting means of a continuous industrial process where a nonorderly shutdown will introduce additional or increased hazard. Exception 2 states that it does not apply

to fire pumps. Specific settings and testing procedures for this ground-fault protection of equipment are covered in Section 230-95. Additional ground-fault protection of feeders where ground-fault protection has not been provided on the service is covered in both Articles 215 and 240.

Example: Where a steel conduit contains either 3/0 AWG THWN insulated copper conductors with two of the 3/0 AWG conductors being in parallel for each phase, and two 3/0 AWG conductors being in parallel for the neutral, (a) would this installation be permitted to be terminated into a 400-ampere main service breaker where the system voltage is 120/208? (The expected loads are evenly divided between fluorescent lighting, data processing equipment, and air-conditioning equipment.) (b) If not, what would be the maximum ampere rating for this service?

Answer: 8 No. 3/0 THWN conductors in a single conduit, Appendix C, Table C8, would require a 3-inch rigid metal conduit.

Table 310-16 3/0 THWN has an ampacity of 200 amperes (from 75°C column). Where the number of current-carrying conductors exceed three, the allowable ampacity shall be reduced as shown in the table with seven to nine conductors, a derating of 70% shall be applied. See Table 310-15(b)(2).

Two conductors per phase:

$$2 \times 200 \times 70\% = 280 \text{ amperes}$$

The next higher overcurrent device per Section 240-3 would be 300 amperes. The service would be a 300-ampere service.

The 400 amperes overcurrent device would require 2-350 kcmil per phase and with eight No. 350 THWN would require the rigid metal conduit to be increased to 4 inches. For conductor termination see Section 110-14(c).

For services exceeding 600 volts, nominal, Part H must be read and adhered to carefully. The definition of service point formerly appeared in this section. The several examples below will help clarify the application of *NEC®* Article 230.

Example: A 3-inch steel intermediate metal conduit (IMC) contains a 120/208 3-phase service conductor consisting of eight size 3/0 THWN copper conductors. Each phase consists of two conductors per phase, and the full-size, neutral (grounded conductor) consists of two conductors per phase terminating in a 400-ampere service disconnecting means consisting of an inverse time circuit breaker. The data processing, fluorescent lighting, and air-conditioning load is evenly divided. Does this installation meet Code?

Step 1. Chapter 9 Table 1 permits a 40% fill of this conduit. For 3/0 type THWN conductors, Appendix C Table C4A would allow eight 3/0 THWN conductors in a 2½ inch conduit. Therefore, the 3-inch IMC conduit is acceptable.

NOTE: Due to the change in the tables in Chapter 9 and the addition of Appendix C in the 1999 *NEC®*, each type of raceway must be calculated individually. For example, in this problem if nonmetallic conduit schedule 40 were used, a 3-inch would be required. If more than one size conductor was contained within the raceway, then calculations would be necessary using Chapter 9 Tables 1, 4, and 5 for the number of conductors, the specific type and size, and the specific raceway type.

Step 2. The ampacity of the 3/0 THWN conductors is 200 amperes each, based on Table 310-16. However, Table 310-15(b)(2) requires a deration to 70% where eight conductors are in one raceway. So, 200 amperes × 2 = 400 × 70% = 280 ampere ampacity. Section 230-90(a) Exception 2 references Section 240-3(b), (c), and Section 240-6, which permits overload protection of the conductors at the next higher standard size overcurrent device in accordance with Section 240-6.

Thus, the largest size overcurrent device permitted would be a 300-ampere device.

Step 3. Assuming the calculated load required the 400-ampere device, you would be required to either increase the conductor size, which would indubitably require a larger conduit, or run two conduits in parallel with conductors sized to carry the calculated load after derating.

Step 4. A 400-ampere calculated load would require a 3½ inch IMC and a 4-inch PVC schedule 40, nonmetallic raceway to contain the eight required 350-kcmil type THWN conductors.

Answer: Table 310-16 and Table 310-15(b)(2); $310 \times 2 = 620 \times 70\% = 434$

SERVICES OVER 600 VOLTS

Services over 600 volts nominal introduce an additional level of hazard. Article 230 Part H covers the requirements for services exceeding 600 volts nominal, which modify or amend the rest of Article 230. All of 230 is applicable and, in addition, Part H must be followed. Clearance requirements for over-600-volt services can be governed by ANSI C2, the *National Electrical Safety Code*, as noted in the fine print note in Section 230-200. The service-entrance conductors to buildings or enclosures shall not be smaller than No. 6 unless in cable, and in cable not smaller than No. 8, and installed by the specific methods listed in Section 230-202(b). It should be noted that Article 300, Part B and Article 110, Part B also will need to be regularly referenced in making installations over 600 volts. The requirements for support guarding and draining cables are covered in this section. Warning signs with the words "Danger. High Voltage. Keep Out." shall be posted where unauthorized people come in contact with energized parts. The disconnecting means must comply with Section 230-70 or 230-208(b) and shall simultaneously disconnect all underground conductors and shall have fault closing rating of not less than the maximum short circuit available in supply terminals, except where used switches or separate mining fuses are installed, the fuse characteristic shall be permitted to contribute to the fault closing rating of the disconnecting means. These requirements for over-600-volt services create the need for your studies to take you into new areas of the Code, such as Part B of Article 100, which covers the general requirements for over 600 volts nominal; Part B of Article 100, which gives you the unique definitions for those circuits; Part I of Article 240; Part K of Article 250 for grounding of system; and Part B of Article 300; and the requirements of Article 490. You must be aware that these additional requirements amend or augment the primary requirement's in each article. In most instances, those requirements also apply to over-600-volt nominal services. Therefore, as you study this section, familiarize yourself with these new parts as they apply to your studies.

CHAPTER 7 QUESTION REVIEW

> **Each lesson is designed purposely to require the student to apply the entire *NEC*® text, not specific chapters or articles. It has been found that when studying for a timed, open-book examination the student must gain proficiency in the Table of Contents, the Index, and the ability to move quickly from cover to cover to find the correct answers to each question in a timely fashion.**

1. Where a circuit breaker is provided as the short-circuit protective device for service-entrance conductors exceeding 600 volts, the breaker shall have a trip setting of not more than _____ times the ampacity of the conductors.

   Answer: _____

   _____

   _____

   _____

   Reference: _____

2. Where a group of buildings are served by electrical feeders from a single service-drop or electrical service point in another building, is a disconnecting means required at each of the other buildings where the feeders terminate or can the feeders terminate in main lug-only panels?

   Answer: _____

   _____

   _____

   _____

   Reference: _____

3. Is there a limit to the number of overcurrent devices that can be used on the secondary side of a transformer if the total of the device ratings do not exceed the allowed value for a single secondary overcurrent device?

Answer: _____

_____

_____

_____

Reference: _____

4. Does it make a difference which opening in the service head that the service-entrance conductors are brought out of? If so, why?

Answer: _____

_____

_____

_____

Reference: _____

5. If the clearances required by Section 110-16 can be maintained in front of a panelboard or a service panel to be installed in a bathroom in a dwelling, would this installation be permitted by the *NEC*®?

Answer: _____

_____

_____

_____

Reference: _____

6. When service-entrance cable is used within a dwelling for branch-circuit wiring, is the SEC-type cable considered the same as Type NM nonmetallic sheath cable?

Answer: _____

_____

_____

_____

Reference: _____

7. Can a 120/240-volt single-phase, 3-wire service be located closer than 3 feet to a window of a dwelling?

Answer: _____

_____

_____

_____

Reference: _____

8. The Code generally prohibits splicing a grounding-electrode conductor. (a) Is this conductor permitted to be tapped? (b) If so, how are these taps sized?

   Answers: _____

   _____

   _____

   _____

   Reference: _____

9. Circuit breakers are required to open all ungrounded conductors of a circuit simultaneously. Obviously, fuses cannot be required to open all ungrounded conductors of a circuit simultaneously. Why the difference in the requirements?

   Answer: _____

   _____

   _____

   _____

   Reference: _____

10. What is a separately derived system? Where is it mentioned in the Code?

    Answer: _____

    _____

    _____

    _____

    Reference: _____

11. How many overcurrent devices are permitted to be used on the secondary of a transformer if the total of the device ratings do not exceed the allowed value for a single overcurrent device?

    Answer: _____

    _____

    _____

    _____

    Reference: _____

12. Is it necessary to identify the higher voltage-to-ground phase (hi-leg) at the disconnect of a 3-phase motor where the service available is 120/240-volt, 3-phase, 4-wire?

    Answer: _____

    _____

    _____

    _____

    Reference: _____

13. A service panel is securely bolted to the metal frame of a building that is effectively grounded, and the panelboard's main bonding jumper is properly installed. Is this sufficient to ground the panelboard to the metal building grounding electrode as required in Section 250-81(b)?

    Answer: _____

    _____

    _____

    _____

    Reference: _____

14. What is the maximum permitted contraction or expansion for a run of Schedule 40 PVC conduit before the installation of an expansion coupling is required?

    Answer: _____

    _____

    _____

    _____

    Reference: _____

15. On a recent installation, the plans called for the building steel to be used as a grounding electrode if effectively grounded. What does this term mean, and where is it used in the *NEC®*?

    Answer: _____

    _____

    _____

    _____

    Reference: _____

16. What article of the *National Electrical Code®* covers Romex?

    Answer: _____

    _____

    _____

    _____

    Reference: _____

17. What is the maximum size flexible metallic tubing permitted?

    Answer: _____

    _____

    _____

    _____

    Reference: _____

18. Which article of the *National Electrical Code*® contains the requirements for fusible safety switches and other type snap switches?

Answer: _____

_____

_____

_____

Reference:_____

19. In making an installation in a residential garage, you have determined that physical protection is needed for the nonmetallic sheath cable and that it must be installed in steel conduit. How do you size the conduit for this nonmetallic sheath cable?

Answer: _____

_____

_____

_____

Reference:_____

20. In making an installation for a large parking garage adjacent to a hotel building, which article of the *National Electrical Code*® covers these wiring methods?

Answer: _____

_____

_____

_____

Reference:_____

Chapter Eight

OVERCURRENT PROTECTION

NEC® Article 240 provides the general requirements for overcurrent protection and overcurrent protective devices not more than 600 volts, nominal. Part I of *NEC®* Article 240 covers the overcurrent protection for over 600 volts, nominal. The overcurrent protection for conductors and equipment is provided to open the circuit if the current reaches a value that will cause an excessive or dangerous temperature in conductors or conductor insulation. Sections 110-9 and 110-10 cover the requirements for the interrupting capacity and protection against fault currents. (See Figure 8–1.) Article 240 covers the general requirements. The specific overcurrent requirements for equipment can be found in each individual article as it pertains to that specific equipment or circuitry; for example, the overcurrent protection for air-conditioning and refrigeration equipment is covered in Article 440 and appliances in Article 422. A listing of the specific overcurrent requirements can be found in Section 240-2. Section 240-3 outlines the protection of conductors. This section provides the rules for the many different applications throughout the Code. Section 240-4 covers the requirements for protecting flexible cords and fixture wires, and standard overcurrent devices are listed in Section 240-6 for fuses and fixed-trip circuit breakers. See Figure 8–2 and Figure 8–3 for examples. Adjustable trip circuit breakers are described and covered in Section 240-6(b). Part B of Article 240 specifies the location that the overcurrent device must be placed in the circuit. Part C specifies the enclosure required for enclosing the overcurrent devices. Part D covers the disconnecting requirements for disconnecting the overcurrent devices so they are accessible to the maintenance personnel performing work on those devices. Parts E and F cover fuses, fuseholders, and adapters for both plug- and cartridge-type fuses, and circuit breakers are covered in Part G. Article 240 Part H covers supervised industrial installations. Part I covers the overcurrent protection for over 600 volts, nominal, and states that feeders shall have short-circuit protection devices in each ungrounded conductor or comply with Section 230-208(a)

Figure 8–1 Panelboards. *(Courtesy of Square D Company)*

Figure 8–2 Types IFL, IKL, and ILL "1 LIMITER" molded case current-limiting circuit breakers. *(Courtesy of Square D Company)*

Figure 8–3 Class T current-limiting, fast-acting fuse; 200,000-ampere interrupting rating, 300 and 600 volts. Has little time delay. Generally used for protection of circuit-breaker panels and for circuits that do not have high-inrush loads, such as motors. When used on high-inrush loads, generally size at 300% so as to be able to override the momentary inrush current. These fuses have different dimensions compared with ordinary fuses. They will not fit into switches made for other classes of fuses, nor will other classes of fuses fit into a Class T disconnect switch. *(Courtesy of Bussmann, Cooper Industries)*

and (b). The protective device(s) shall be capable of detecting and interrupting all values of current that can occur at their location in excess of their trip setting or melting point. In no case shall the fuse rating exceed three times the long-time trip element setting of the breaker, or six times the ampacity of the conductor. Branch-circuit requirements for over-600-volt circuits are covered in Section 240-100.

Example: What is the maximum size time-delay fuse permitted to protect a 1-horsepower, 3-phase, 230-volt squirrel cage motor with a nameplate of 3.3 amperes?

Step 1. Section 430-6. Use the tables and not the nameplate amperes.

Step 2. Table 430-150. One horsepower is 3.6 amperes.

Step 3. Section 430-52(a) Exception 1. (Where the standard overcurrent device size does not correspond as permitted by Table 430-152, the next higher is permitted.)

Step 4. Table 430-152 permits 175% for a time-delay fuse. 3.6 amperes × 175% = 6.3 amperes.

Step 5. Section 240-6. The next higher standard fuse is 10 amperes. Therefore, a 10-ampere time-delay fuse would be permitted.

Answer: 10 amperes

Example: A 120/208 3-phase lighting and branch-circuit panelboard is to be fed with 23-foot long copper tap conductors tapped from three 350 kcmil THWN copper conductors run in parallel protected by an 800-ampere circuit breaker. The panelboard supplies a continuous-duty lighting load of 135 amperes per phase; one hundred five 120-volt receptacle outlets; four 3-horsepower, 3-phase squirrel-cage motors; and one 25-horsepower, 3-phase, 5-minute-rated, short-time-duty squirrel-cage motor. What are the minimum size tap conductors required?

Step 1. Index "Taps." See *Feeders, Feeder taps,* Sections 240-21 and 430-28.

Step 2. Although motors are a portion of the intended load, these taps are to feed a panelboard with mixed loads.

Step 3. See Section 240-21(c). Since 23 feet is more than 10 and less than 25, we must use the 25-foot rule.

Step 4. The ampacity of the tap conductors must be a minimum of one-third the rating of the overcurrent device protecting the circuit. Therefore, 800 × ⅓ = 267 amperes. Reference Table 310-16. THWN Copper 75°C would require **a minimum tap conductor size of 300 kcmil**. Now, calculations are necessary to see if the load exceeds the minimum. If it does, then the larger of the two would be required.

Continuous duty lighting 135 × 125%   = 168.75
  Section 220-10(b)

105 receptacles @ 180 VA = 18,900 VA
  @ 10,000 + (8900 × 50%) =
  14,450 ÷ 120   = 120.4
    Section 220-4(c)(6) Table 220-13

Three 3 hp motors @ 10.6   = 31.8
  Section 430-6, Table 430-150

One 25 hp motor 5-minute short-time
  duty. 74.8 × 110%   = 82.28
    Section 430-24 Exception 1,
    430-22 Exception 1, and
    Table 430-150

Round to the nearest whole number = 403

Answer: The minimum size is not adequate. Therefore, 600 kcmil per phase with a 420 ampacity would be required (Section 310-16).

Example: What is the minimum size copper THW feeder required for a single-phase 120/240-volt office building feeder with the maximum unbalanced load of 125 amperes of continuous-duty fluorescent lighting on each ungrounded conductor?

Step 1. Index "Feeder Calculation of Loads." Sections 220-10(a) and 220-30, and Chapter 9 Part B.

Step 2. Section 220-10(a) requires that continuous-duty loads be calculated at the load plus 25% for sizing the overcurrent device.

Step 3. 125 × 125% = 146.25 amperes. See Section 240-3(b) = 150-ampere overcurrent device.

Step 4. Section 310-15, Table 310-16 Ampacity tables.

Answer: 1/0 copper-type THW conductors

Example: Provide the fuse sizes for a 10-horsepower, 230-volt, 3-phase motor code lettered G, full voltage start with a service factor of 1.15.

(A) The fuse sizes using time-delay fuses for the branch-circuit protection is calculated as follows: 28 × 1.75 = 49 amperes. (Section 430-52 requires that you round down to the next lower standard size. See Section 240-6.) Therefore, install 45-ampere time-delay fuses. If the 45-ampere time-delay fuses are not sufficient to carry the load or to start the motor, a larger size time-delay fuse may be installed, but not to exceed 225%.

(B) Using time-delay fuses for motor overload protection, the size will be as follows: 28 × 1.25 = 35-ampere time-delay fuses.

(C) Using nontime-delay fuses, sizes for branch-circuit and ground-fault protection are calculated as follows: 28 × 3 = 84 amperes. Round down to 80-ampere nontime-delay fuses as explained in the example above. This would permit a maximum size of 28 × 4 = 112. Install 110-ampere nontime-delay fuses. References for making these calculations are Tables 430-150 and 152, Sections 240-6, 430-32, and 430-52.

Example: To calculate the available short-circuit current at a panelboard located 20 feet away from a transformer when the transformers are 500 kcmil copper in steel conduit. The (c) value for the conductors is 22,185. The transformer is marked 300 kVA, 208/120 volts, 3-phase, 4-wire. The transformer impedance is 2%. Consider the source to have an infinite amount of fault current available (often referred to as "infinite primary"). Find the transformer's full-load secondary current,

$$I_{fla} = \frac{kVA \times 1000}{E \times 1.73} = \frac{300 \times 1000}{208 \times 1.73} = 834 \text{ amperes}$$

(2) To find the transformer SCA:
multiplier = 100/2 = 50.

Transformer SCA =

$I_{fla} \times$ multiplier = 834 × 50 = 41,700 amperes

(3) To find the (F) factor:

$$F = \frac{1.73 \times L \times I}{C \times E_{L-L}} = \frac{1.73 \times 20 \times 41,700}{22,185 \times 208} = 0.3127$$

(4) To find the (M) multiplier:

$$M = \frac{1}{1 + F} = \frac{1}{1 + 0.3127} = 0.76$$

(5) To find the short-circuit current of the panelboard:

$I_{sca} =$ transformer$_{sca} \times$ (M) = 41,700 × 0.76 = 31,692 amperes

The fault current available at the panelboard where it is located 20 feet away from the transformer is equal to 31,692 amperes.

CHAPTER 8 QUESTION REVIEW

> Each lesson is designed purposely to require the student to apply the entire *NEC®* text, not specific chapters or articles. It has been found that when studying for a timed, open-book examination the student must gain proficiency in the Table of Contents, the Index, and the ability to move quickly from cover to cover to find the correct answers to each question in a timely fashion.

1. Does the Code permit the use of two single-pole circuit breakers in a panelboard to serve a line-to-line connected load such as a household electrical range rated at 120/240 volts or a hot water heater rated at 240 volts?

    Answer: _____

    _____

    _____

    _____

    Reference: _____

2. Where outdoor conductors are tapped and these tapped conductors terminate into a single overcurrent device designed to limit the load so as not to exceed the ampacity of the tap conductors, how long can these tap conductors be at the maximum length?

    Answer: _____

    _____

    _____

    _____

    Reference: _____

3. Is it permissible to use 300-volt cartridge-type fuses and fuseholders to protect a multiwire, 4-wire circuit such as an emergency circuit in a food store where the electrical service is 277/480-volt, 3-phase, 4-wire?

    Answer: _____

    _____

    _____

    _____

    Reference: _____

4. Service-entrance conductors are required to be protected at their rated ampacity with exception, the exception being the next standard size overcurrent device when the service conductors are protected by a single device. When are multiple overcurrent devices supplied by a service allowed to exceed the ampacity rating of the service-entrance conductors?

    Answer: _____

    _____

    _____

    _____

    Reference: _____

5. Fixed electric space heating loads shall be computed at _____ % of the total connected load, generally.

Answer: _____

_____

_____

_____

Reference: _____

6. A motor is to be installed in a Class I, Division 2 location. It is to be connected with 3 feet of $\frac{3}{4}$ inch liquidtight flexible metal conduit. Is it permissible to use the liquidtight flexible metal conduit as the grounding path for this installation?

Answer: _____

_____

_____

_____

Reference: _____

7. Are feeder conductors required to be rated for 125% of the continuous load plus the noncontinuous load?

Answer: _____

_____

_____

_____

Reference: _____

8. How many duplex receptacles can be installed on a 20-ampere branch-circuit in a dwelling?

Answer: _____

_____

_____

_____

Reference: _____

9. Table 310-16 lists the allowable ampacity of No. 12 THWN conductors as 25 amperes. Can 16 general-use receptacles in an office building be connected to a 20-ampere circuit?

    Answer: _____

    _____

    _____

    _____

    Reference: _____

10. When an electric heat pump is installed with backup resistance heat in a building, (a) could one of the loads ever be considered dissimilar for the purposes of calculating the service loads? (b) Would the heat pump load have to be calculated at 125% of the service sizing?

    Answer: _____

    _____

    _____

    _____

    Reference: _____

11. Are all 20-ampere residential underground circuits required to have GFCI protection?

    Answer: _____

    _____

    _____

    _____

    Reference: _____

12. Can two circuit breakers mounted adjacent to each other in a panelboard with handle ties be used in place of a double-pole breaker to supply a 240-volt electric baseboard heater?

    Answer: _____

    _____

    _____

    _____

    Reference: _____

13. Are water heaters considered to be a continuous load so that a 125% load is required for calculating feeder and service on an installation?

Answer: _____

_____

_____

_____

Reference: _____

14. Is it permitted to plug a microwave oven with a nameplate rating of 13 amperes into a 15-ampere receptacle protected on a 20-ampere circuit?

Answer: _____

_____

_____

_____

Reference: _____

15. Can a 1,400 VA load, a cord-, and plug-connected load, be supplied by a 240-volt circuit in a residence?

Answer: _____

_____

_____

_____

Reference: _____

16. Three receptacles on a single yoke or strap are to be installed in a commercial or industrial facility. Are these outlets to be calculated at 180 VA or 540 VA?

Answer: _____

_____

_____

_____

Reference: _____

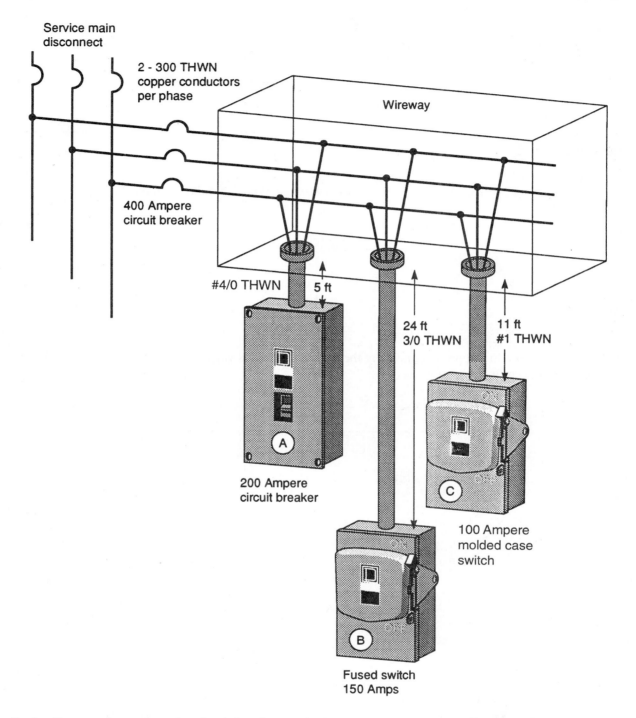

Service main disconnect

2 - 300 THWN copper conductors per phase

Wireway

400 Ampere circuit breaker

#4/0 THWN

5 ft

24 ft 3/0 THWN

11 ft #1 THWN

200 Ampere circuit breaker

A

100 Ampere molded case switch

C

Fused switch 150 Amps

B

17. In the diagram above, does the circuit breaker marked A comply with the *NEC®*? If so, which section?

Answer: _____

_____

_____

_____

Reference: _____

18. In the diagram on page 103, does the switch B comply with the *NEC*®? If so, which section?

Answer: _____

_____

_____

_____

Reference: _____

19. In the diagram on page 103, does the molded case switch marked C comply with the *NEC*®?

Answer: _____

_____

_____

_____

Reference: _____

20. *NEC*® requires overcurrent protection where the last 50 feet of busway is reduced in size where the installation is other than industrial. (True or False)

Answer: _____

_____

_____

_____

Reference: _____

Chapter Nine

GROUNDING

NEC® Article 250 covers the general requirements and many specific requirements for grounding such as systems, circuits, and equipment required, permitted, and/or not permitted to be grounded, circuit conductors to be grounded on grounded systems, location of the grounding connections, type and sizes of grounding and bonding conductors and electrodes, the methods of grounding and bonding, and the conditions under which guards, isola-

tion, or insulation can substitute for grounding. Other applicable articles applying to particular cases of installations of conductors and equipment are found in *NEC*® Section 250-4. (See Figure 9–1.)

The rules and performance requirements of grounding and bonding are clearly stated in Section 250-2 (a) through (d). Article 250 has been formatted to the proper *NEC*® style and renumbered. The majority of exceptions and fine print notes have been rewritten into positive language or deleted. The specific requirements

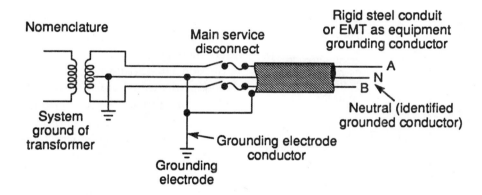

Requirements for grounding electrode conductor, grounded system

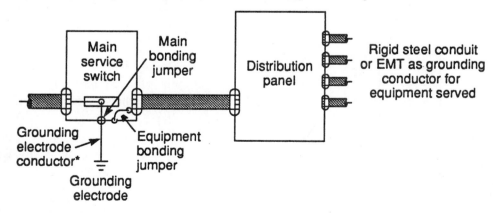

*Grounds both the identified grounded conductor and the equipment grounding conductor. It is connected to the grounded metalic cold water piping on premises, or other available grounding electrode, such as metal building frame if effectively grounded, or a made electrode.

Figure 9–1 Grounding principles. *(Courtesy of American Iron & Steel Institute)*

have been grouped into Parts A through K and they are titled as follows:

- Part A—General
- Part B—Circuit and System Grounding
- Part C—Grounding Electrode System and Grounding Electrode Conductor
- Part D—Enclosure, Raceway, and Service Cable Grounding
- Part E—Bonding
- Part F—Equipment Grounding and Equipment Grounding Conductors
- Part G—Methods of Equipment Grounding
- Part H—Direct-Current Systems
- Part J—Instruments, Meters, and Relays
- Part K—Grounding of Systems and Circuits of 1 kV and Over (High Voltage)

NEC® Part H for direct-current systems, Part B for alternating-current circuits and systems (see Figure 9–2),

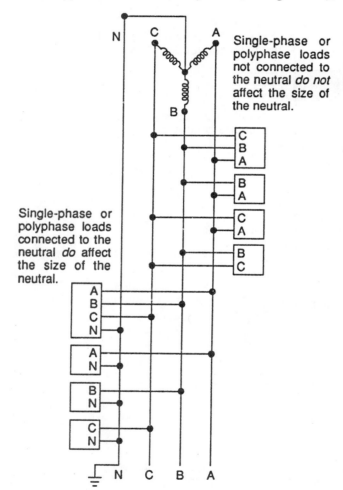

Figure 9–2 Size of a neutral conductor as in *NEC*® Section 220-4(e). *(Courtesy of American Iron & Steel Institute)*

and Section 250-34 for portable and vehicle mounted generators outline the conditions for which these systems and circuits are to be grounded or can be grounded. Section 250-22 covers the circuits not to be grounded. Sections 250-24 and 250-32 cover the requirements necessary for grounding buildings, single or locations with two or more buildings fed from a common service.

Example: The following example illustrates how the size of the grounded (neutral) wire may be selected. A new 3-phase, 4-wire feeder is connected in an existing junction box to three single-phase plus grounded circuits, each consisting of 2-phase conductors and a grounded conductor. Two of these circuits are in one conduit; the third is in another. The single-phase circuit conductors are all #2 THW copper. Determine the ampacity of each single-phase circuit and the required size of new service if electric discharge lighting ballasts are in the circuits but the load is considered noncontinuous.

Step 1. Neutrals in these circuits must be counted as current-carrying conductors. (*NEC*® Section 310-15(b)(4)(a).) The ampacity of #2 THW conductors from Table 310-16 is 115 amperes.

Step 2. The circuit ampacity in the conduit with the single circuit, derating is not required for only three conductors. (*NEC*® Table 310-15(b)(2)(a).) Thus, the allowable ampacity is 115 amperes.

Step 3. In the conduit with two circuits, an 80% derating factor must be used, because all six wires are considered as current-carrying conductors. (*NEC*® Table 310-15(b)(2)(a).) Each of these two circuits has a design ampacity of 80% × 115 = 92.

Step 4. The sum of the ampacities of all three circuits is the required ampacity of the service conductors and the service neutral (Section 220-22).

Step 5. The value is 115 + 92 + 92 = 299. Each conductor and neutral requires not less than a 350 kcmil 75°C conductor ampacity. Reference Article 310, Table 310-16.

Section 250-26 outlines the requirements for grounding a separately derived alternating-current system. Article 250, Part D covers the requirements for grounding the enclosures, Part E the equipment grounding requirements, Part F covers the methods for grounding, and Part G the bonding requirements. (See Figure 9–3 and Figure 9–4.)

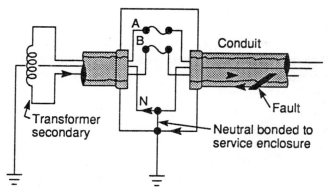

Figure 9–3 Bonding at service disconnect enclosure as required in *NEC*® Section 250-71. *(Courtesy of American Iron & Steel Institute)*

NEC, Article 250, Part E

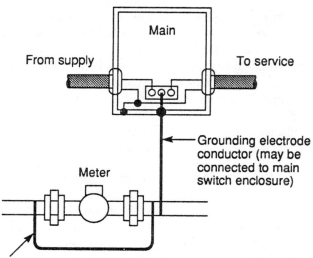

Bonding jumper around water meter and other fittings in water piping likely to be disconnected, such as unions, where grounding connection is not on street side of water meter.

Figure 9–4 Bonding jumpers as in *NEC*® Article 250, Part E. *(Courtesy of American Iron & Steel Institute)*

Part G clearly delineates between service equipment and other-than-service equipment.

Caution: It is extremely important for only one grounding system to exist in an installation because multiple-separated grounds can create unwanted ground loops that may create errors in data transmission and malfunctions. Where a ground loop exists, know techniques will solve the problem. Although the single-point ground is preferred, it may not always be feasible where long distances and very high frequencies are present.

Part H clearly describes the grounding electrode systems (see Figure 9–4), and Part J the grounding conductors, types, sizes, and enclosures for those conductors. Most important are Table 250-66 for sizing the grounding-electrode conductor for an AC system, and the Exceptions to Section 250-66 as they apply to that table. Table 250-122 lists the minimum equipment grounding conductor for grounding raceway and equipment. Section 250-122 and Exceptions provide rules for applying that table. Part J covers instrument transformers, and so on. Part K covers systems and circuits of 1 kV (1,000 volts) or over (high voltage). Where high-voltage systems are grounded, they must comply with all applicable provisions of all Article 250, Part K, which supplements or modifies Parts A through J. Sections 250-80 and 250-86 clarify a common misconception of adding an insulated equipment grounding conductor with the circuit conductors from the source, service, or subpanel to the equipment being served and terminating on the ends relieves one responsibility of installing the metallic raceways as a complete system. All metal raceway enclosures must be bonded to the equipment grounding conductor to ensure that no potential exists on that equipment, and if one of the conductors fault to the raceway would occur, that the overcurrent device will trip. The insulated conductor when installed is redundant and not required. (The redundant conductor will only carry about 3% to 5% of the current flowing, while the metal raceway will carry 95% to 97% of the current flowing.) Article 250 describes the general requirements for equipment grounding and provides a list of specific items, both the electrical and nonelectrical equipment fastened in place, and cord-connected equipment in residential occupancies and other than residential occupancies. The list is specific. Article 250 also covers the limited requirement for spacing lightning rods, Section 250-106 supplements the requirements in NFPA 780, which contains detailed information on grounding lightning protection systems. The types of grounding-electrodes and details for the grounding-electrode system are found in Sections 250-50 through 250-60. Related to those sections, you also should read Section 250-104, which is the bonding requirements for piping systems and exposed building steel, (a) covering metal water piping and (b) other metal piping, which provide additional safety to the system. As you read Section 250-50 and Exceptions, you will see that all available grounding electrodes on each premise must be used and bonded together. One electrode system must be provided where the building contains others, such as building steel, metal water pipes, and a concrete-encased electrode; all must be used and bonded together. Read these sections carefully before applying them. Section 250-58 requires common grounding electrodes and states, *Where an ac system is connected to a grounding electrode in or at a building as specified in Sections 250-24 and 250-32, the same electrode shall be used to ground conductor enclosures and equipment in or on that building. Where separate services supply a building and are required to be connected to a grounding electrode, the same grounding electrode shall be used. Two or more grounding electrodes that are effectively bonded together shall be considered as a single grounding-electrode system in this sense.**

Example: A dairy barn is fed underground from the farmhouse with a 225-ampere overcurrent device using 4/0 copper phase conductors and a No. 1 (neutral) grounded conductor. What size equipment grounding conductor is required for this installation?

Step 1. Index "Grounding, Separate buildings," Section 250-32. Also, "Grounding Conductors, Sizes," Table 250-122.

Step 2. Section 250-32 requires an insulated or covered equipment grounding conductor where livestock is housed. Also, Section 547-8(a) contains similar requirements.

Step 3. Table 250-122. For the minimum size permitted for a 225-ampere overcurrent device, the table lists only a 200 and a 300. Therefore, the 300 would apply, because the 200 would not be adequate.

Answer: Number 4 copper

When sizing the equipment grounding conductor in accordance with Table 250-122, beware this is the minimum size conductor allowed but may not be large enough to comply with other applicable sections in Article 250, such as Section 250-2(d) that states, *The path to ground from circuits, equipment, and metal enclosures for conductors shall: (1) be permanent and continuous; (2) have capacity to conduct safely any fault current likely to be imposed on it; and (3) have sufficiently low impedance to limit the voltage to ground and to facilitate the operation of the circuit protective devices in the circuit. The earth shall not be used as the* *sole equipment grounding conductor.** (This requirement clarifies that driving a grounding electrode at remote equipment is not an acceptable means for grounding.) An equipment grounding conductor must be run with the other conductors as stated in Section 300-3(b) from the supply to the load and sized in accordance with Table 250-122 or larger to facilitate the operation of the overcurrent device protecting that circuit. You will find after studying Article 250 that grounding is one of the most interesting sections of the *NEC*®, and is written for easy understanding. However, careful study and application are necessary to accomplish a safe design and/or installation of the electrical system.

Example: The following example illustrates how the size of the grounded (neutral) wire may be selected. A new 3-phase, 4-wire feeder is connected in an existing junction box to three single-phase plus grounded circuits, each consisting of 2-phase conductors and a grounded conductor. Two of these circuits are in one conduit, the third is in another. The single-phase circuit conductors are all No. 2 THW copper. Determine the ampacity of each single-phase circuit and the required size of new service if electric discharge lighting ballasts are in the circuits, but the load is considered noncontinuous.

Solution: Neutrals in these circuits must be counted as current-carrying conductors. (*NEC*® Section 310-15(b)(4)(a).) The ampacity of No. 2 THW conductors from Table 310-16 is 115 amperes. The circuit ampacity in the conduit with the single circuit, and no derating is required for only three conductors. Thus, the ampacity equals 115 amperes. In the conduit with two circuits, an 80% derating factor must be used, because all six wires are considered as current-carrying conductors. Each of these two circuits has a design ampacity of $80\% \times 115 = 92$. The sum of the ampacities of three circuits is the required ampacity of the service conductors and the service neutral (Section 220-22). This value is $115 + 92 + 92 = 299$. Each conductor and neutral requires not less than a 350 kcmil 75% seat conductor ampacity. Reference Article 310, Table 310-16.

---

* Reprinted with permission from NFPA 70-1999.

CHAPTER 9 QUESTION REVIEW

> **Each lesson is designed purposely to require the student to apply the entire *NEC*® text, not specific chapters or articles. It has been found that when studying for a timed, open-book examination the student must gain proficiency in the Table of Contents, the Index, and the ability to move quickly from cover to cover to find the correct answers to each question in a timely fashion.**

1. Required grounding conductors and bonding jumpers cannot be connected solely by _____ connections.

   Answer: _____

   _____

   _____

   _____

   Reference: _____

2. The conductor permitted to bond together all isolated noncurrent-carrying metal parts of an outline lighting system is at least size _____ AWG copper.

   Answer: _____

   _____

   _____

   _____

   Reference: _____

3. A connection to a concrete-encased, driven or buried grounding electrode shall be _____ .

   Answer: _____

   _____

   _____

   _____

   Reference: _____

4. The *grounded* conductor of an electrical branch-circuit is identified by the color _____ .

   Answer: _____

   _____

   _____

   _____

   Reference: _____

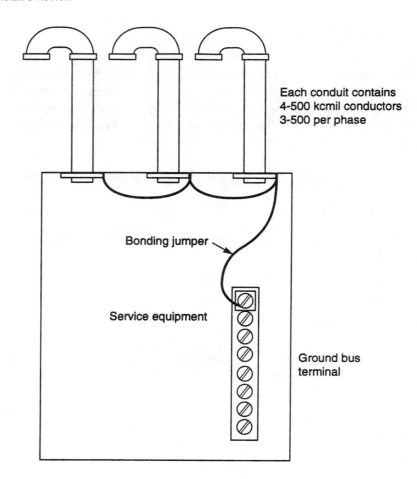

Each conduit contains
4-500 kcmil conductors
3-500 per phase

Bonding jumper

Service equipment

Ground bus
terminal

5. *(Refer to the figure above.)* Each of the supply side EMT conduits contain three 500 kcmil copper THW service conductors in parallel. As shown, only one bonding jumper is used to bond all the conduits to the grounded bus terminal.

This bonding jumper must be at least size _____ AWG copper.

Answer: _____

_____

_____

_____

Reference: _____

6. A rigid metal conduit contains three circuits: two 150 amp, 3-phase circuits, and one 300 amp, single-phase circuit. The load side service equipment bonding jumper for this conduit must be at least size _____ AWG copper.

Answer: _____

_____

_____

_____

Reference: _____

7. Are EMT set screw connectors approved for grounding of service raceways or are jumpers required to bond around them?

   Answer: _____

   _____

   _____

   _____

   Reference: _____

8. Where feeder conductors are paralleled and routed in separate nonmetallic raceways, is an equipment grounding conductor required in each nonmetallic raceway or only in one nonmetallic raceway?

   Answer: _____

   _____

   _____

   _____

   Reference: _____

9. Are submersible deep-well pumps required to have an equipment grounding conductor run with the other conductors and the submersible pump motor housing grounded?

   Answer: _____

   _____

   _____

   _____

   Reference: _____

10. Is it permissible to ground the secondary of a separately derived system to the grounded terminal in the service switchboard instead of the nearest effectively grounded structural steel or nearest water pipe?

    Answer: _____

    _____

    _____

    _____

    Reference: _____

11. Where installing the grounding-electrode conductor in a raceway for physical protection, is the metal raceway required to be bonded to the grounding-electrode conductor? If the answer is yes, is it sufficient to bond it at the service panel only or must bonding be done at all terminations?

    Answer: _____

    _____

    _____

    _____

    Reference: _____

12. Where installing the electrical hookups for RV sites in a recreational vehicle park, can a ground rod be driven at each RV site instead of pulling an equipment grounding conductor with a circuit conductor from the source?

    Answer: _____

    _____

    _____

    _____

    Reference: _____

13. You have recently installed a separately derived system, a dry-type transformer in a commercial building. No effectively grounded structural steel was near the transformer location. The transformer secondary is grounded to a nearby metal water pipe. Is a supplemental grounding electrode required?

    Answer: _____

    _____

    _____

    _____

    Reference: _____

14. When bonding the noncurrent-carrying metal parts of a recreational vehicle, what is the minimum size copper or equivalent bonding conductor?

    Answer: _____

    _____

    _____

    _____

    Reference: _____

15. Does the Code require interior gas piping systems to be bonded to the grounding-electrode system?

    Answer: _____

    _____

    _____

    _____

    Reference: _____

16. Two parallel runs of four 500 kcmil conductors each are run in rigid nonmetallic conduits to feed a 600 amp panel. Is an equipment grounding conductor required in each conduit? If the answer is yes, what size equipment grounding conductor is required in each conduit?

    Answer: _____

    _____

    _____

    _____

    Reference: _____

17. When terminating the equipment grounding conductor in an outlet box supplied by nonmetallic sheath cables, can the equipment grounding conductors be twisted together or must they be connected with a wire connector?

    Answer: _____

    _____

    _____

    _____

    Reference: _____

18. What size aluminum equipment grounding conductor will be required to be run with the circuit conductors fed from a 60-ampere fusible switch?

    Answer: _____

    _____

    _____

    _____

    Reference: _____

19. In a run of conduit supplying a motor, it is necessary to install 4 feet of flexible metal conduit where the raceway leaves the panel to get around a column. When the conduit reaches the motor, an additional 3 feet of flexible metal conduit is employed for convenience and flexibility. Is an equipment grounding conductor required in this metal raceway?

Answer: _____

_____

_____

_____

Reference: _____

20. When making an installation of a high-impedance grounded neutral system, how is the neutral conductor to be sized?

Answer: _____

_____

_____

_____

Reference: _____

Chapter Ten

WIRING METHODS

The general requirements for wiring methods are found in Article 300. (See Figure 10–1, Figure 10–2, and Figure 10–3.) As you may recall in earlier chapters of this book, several times you were told that you must know the contents of Article 300 thoroughly, because it is used in some application for every installation.

The general wiring methods in Article 300 include requirements that apply to the specific wiring methods, such as nonmetallic sheath cable that is covered in Article 336 (see Figure 10–4), as specific support requirements and installation criteria, such as found in Section 300-4(a), (b), (c), and (d) and the Exceptions. These requirements are mandatory and must be adhered to when making an installation of any wiring method and are in Chapter 3 as they apply. Chapter 3 also covers burial depths (in Table 300-5 for conductors under 600 volts) and many conditions that apply to underground installations, such as splices, taps, and the backfill material used so as not to damage the raceway or cable that you have selected as a wiring method for your installation. When a wiring method is installed in earth that is not suitable as physical protection, select fill material should be used, such as sand or soil without rocks. This article also covers such things as seals and protection against corrosion, indoor wet locations, raceways exposed to different temperatures, and expansion joints. This chapter discusses much about the expansion joint requirements for nonmetallic raceways. However, Section 300-7 covers all wiring methods and where they are required to compensate for expansion or contraction.

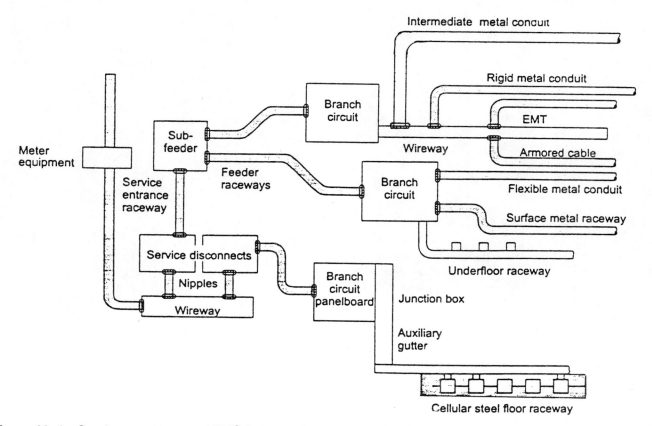

Figure 10–1 Service raceway per *NEC®* Article 230. **Note:** Location of overcurrent devices not shown. *(Courtesy of American Iron & Steel Institute)*

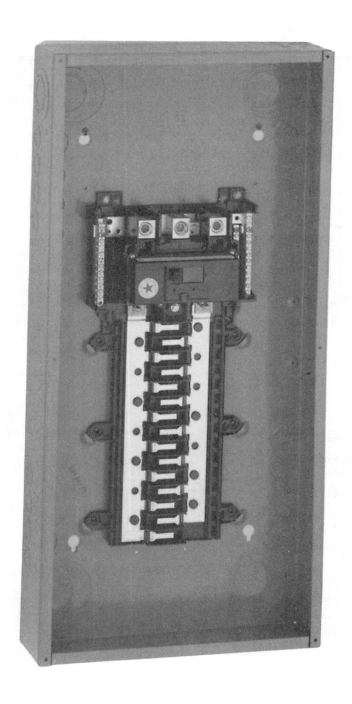

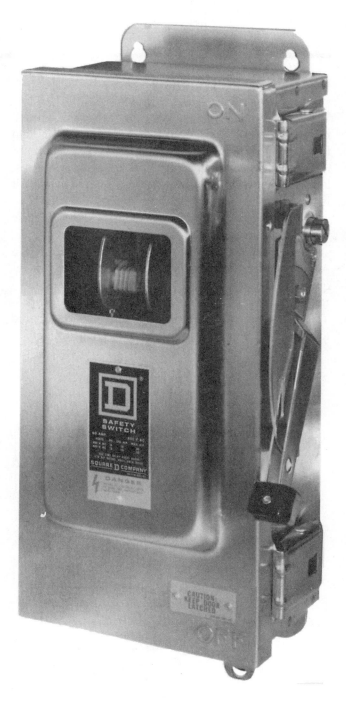

Figure 10–2 Main breaker panelboard interior and enclosure. *(Courtesy of Square D Company)*

Figure 10–3 Typical safety switch with view safety glass. *(Courtesy of Square D Company)*

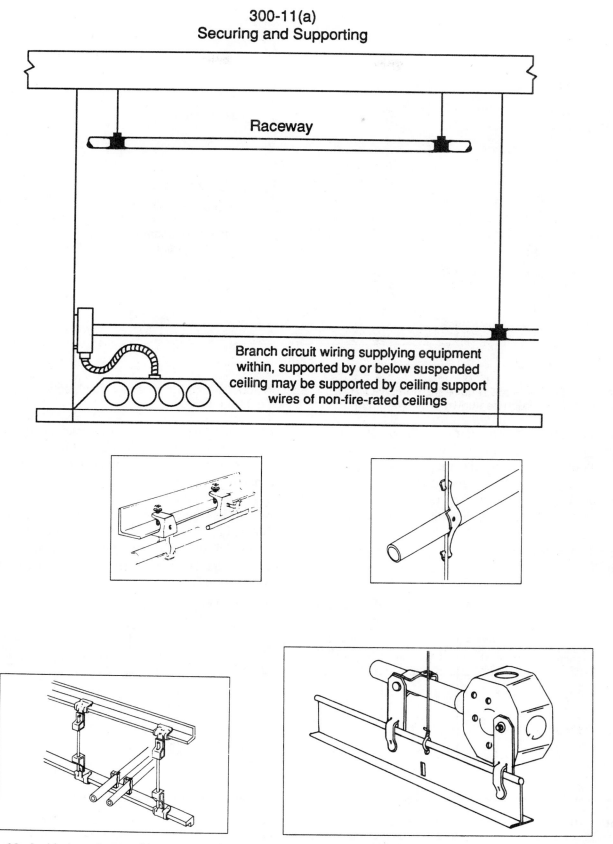

300-11(a)
Securing and Supporting

Raceway

Branch circuit wiring supplying equipment
within, supported by or below suspended
ceiling may be supported by ceiling support
wires of non-fire-rated ceilings

Figure 10–4 Various types of hangers and supporting methods and hardware for raceways. *(Courtesy of B-Line Systems, Inc.)*

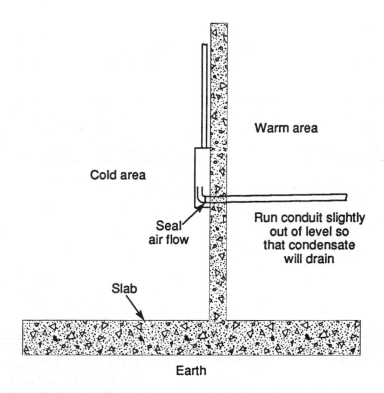

Warm area

Cold area

Seal
air flow

Run conduit slightly
out of level so
that condensate
will drain

Slab

Earth

Figure 10–5   Seals are required to comply with *NEC*® Section 300-7 where the raceway is subjected to more than one temperature such as shown. The system enters the building from outside.

(See Figure 10–5, Figure 10–6, and Figure 10–7.) The electrical and mechanical continuity for metallic systems, both the raceway and the enclosures, Section 300-11 requires that these methods, raceways, cable assemblies, boxes, cabinets, and fittings all be securely fastened where supported on support wires above suspended ceilings.

For branch-circuit wiring, read this section carefully. The wiring systems for feeders or services are not permitted to be secured in this manner; only certain branch-circuits are permitted. Specifically, the junction box out of the box section in Article 370 places more stringent requirements for supporting these box systems. All requirements must be followed. Article 300 also tells us where a box or fitting is required and where a box only is permitted. Article 300-15(a) states that all fittings and connectors must be designed for the specific wiring method for which they are used and listed for that

Table 10.   Expansion Characteristics of PVC Rigid Nonmetallic Conduit
Coefficient of Thermal Expansion = 3.38 × 10⁻⁵ in./in./°F

Temperature Change in Degrees F	Length Change in Inches per 100 ft. of PVC Conduit	Temperature Change in Degrees F	Length Change in Inches per 100 ft. of PVC Conduit	Temperature Change in Degrees F	Length Change in Inches per 100 ft. of PVC Conduit	Temperature Change in Degrees F	Length Change in Inches per 100 ft. of PVC Conduit
5	0.2	55	2.2	105	4.2	155	6.3
10	0.4	60	2.4	110	4.5	160	6.5
15	0.6	65	2.6	115	4.7	165	6.7
20	0.8	70	2.8	120	4.9	170	6.9
25	1.0	75	3.0	125	5.1	175	7.1
30	1.2	80	3.2	130	5.3	180	7.3
35	1.4	85	3.4	135	5.5	185	7.5
40	1.6	90	3.6	140	5.7	190	7.7
45	1.8	95	3.8	145	5.9	195	7.9
50	2.0	100	4.1	150	6.1	200	8.1

Figure 10–6   (Reprinted with permission from NFPA 70-1999, the *National Electrical Code*®, Copyright © 1998, National Fire Protection Association, Quincy, MA 02269. This reprinted material is not the complete and official position of the National Fire Protection Association on the referenced subject, which is represented only by the standard in its entirety.)

Example

380 ft. of conduit is to be installed on the outside of a building exposed to the sun in a single straight run. It is expected that the conduit will vary in temperature from 0°F in the winter to 140°F in the summer (this includes the 30°F for radiant heating from the sun). The installation is to be made at a conduit temperature of 90°F. From the table, a 140°F temperature change will cause a 5.7 in. length change in 100 ft. of conduit. The total change for this example is 5.7″ × 3.8 = 21.67″ which should be rounded to 22″. The number of expansion couplings will be 22 ÷ coupling range (6″ for E945, 2″ for E955). If the E945 coupling is used, the number will be 22 ÷ 6 = 3.67 which should be rounded to 4. The coupling should be placed at 95 ft. intervals (380 ÷ 4). The proper piston setting at the time of installation is calculated as explained above.

$$ 0 = \left[\frac{140 - 90}{140} \right] 6.0 = 2.1 \text{ in.} $$

Insert the piston into the barrel to the maximum depth. Place a mark on the piston at the end of the barrel. To properly set the piston, pull the piston out of the barrel to correspond to the 2.1 in. calculated above.

See the drawing below.

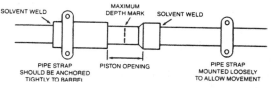

Figure 10–7 Rigid nonmetallic conduit. For steel raceway (21.7 × .1 = 2.17 inches), an expansion fitting may be required. *(Courtesy of Carlon, a Lamson & Sessions Company)*

purpose. Look up the definition for a branch-circuit listed in Article 100 now. An important requirement for all wiring methods is found in Section 300-21. The requirement limits the spread of fire or products of combustion and requires that the wiring system be installed so as to minimize the spread of fire and products of combustion; that all penetrations through floors, walls, and ceilings be sealed in an improved manner to accomplish this purpose. Section 300-22, which are the provisions for the installation and uses of electrical wiring and equipment in ducts, plenums, and other air handling spaces, states these sections are safety concerns in determining the wiring method to be used in electrical installations; these should be studied carefully before making the final selections. Part B of Article 300 is the requirements for installations over 600 volts nominal. Section 300-50 covers the underground installations and references the table and section in Section 300-50, which are the general wiring requirements for installations over 600 volts. Other general articles are found in Chapter 3 relating to wiring methods. Article 305 covers the temporary wiring requirements needed to provide power for the workers making the installation

and constructing the facility during the construction phases of a job. Article 305 also covers other types of temporary wiring, such as temporary wiring concerned with Christmas tree lighting, trade fairs, emergencies and tests, and experimental and development work. Be aware of Article 305. Article 310 is the conductors to be used for general wiring. In Article 310 you will find all requirements related to conductors for general wiring. They do not apply to conductors that form an integral part of equipment such as motors, motor controllers, and similar equipment, or conductors specifically provided for elsewhere in the Code. An example would be Article 400 and Article 402 for flexible cords and fixture wires. Other sections dealing with conductors would be Chapter 7 of the *NEC®* and Chapter 8 of the *NEC®*, which deal with limited voltage wire, such as in Article 725, 760, and so on. Article 310 must be used in every installation. The requirements in 310-1 through 310-15 are general requirements for the use and installation of conductors. The tables apply as applicable.

NOTE: When using the tables, read each heading carefully. Read the columns carefully. Read the ambient temperature correction factors below each column, and remember that the notes (footnotes) under each table are part of those tables and are mandatory requirements. This is important when studying for a test because many test questions will require you to study these additional parts of the table to get the correct answer.

For example, the obelisk note under each table applies to general wiring methods. However, notice that it says unless specifically otherwise provided for or permitted in this Code. Motors, for instance, generally are not required to comply with this obelisk note because Article 430 has specific requirements for branch-circuit conductors for motors. Study these tables carefully. The next important part of Article 310 is Section 310-15(b)(6). Table 310-15(b)(6) has less restrictive ampacity requirements for dwelling services and feeders.

NOTE: These only apply to single-phase, 120/240-volt, 3-wire systems, and do not apply to any other type of system.

Example: *NEC®* Section 318-9(a)(1)

Width selection for cable tray containing 600-volt multiconductor cables, sizes No. 4/0

NEC® Section 318-9(a)(1)

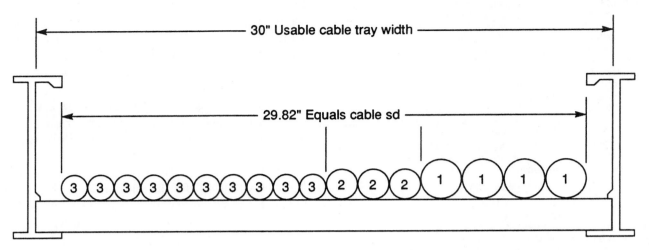

Cross Section of the Cables and the Cable Tray

(Courtesy of B-Line Systems)

AWG and larger only. Cable installation is limited to a single layer. The sum of the cable diameters (Sd) must be equal to or less than the usable cable tray width.

Cable tray width is obtained as follows:

A—Width required for No. 4/0 AWG and larger multiconductor cables:

Item Number	List Cable Sizes	(D) List Cable Outside Diameter	(N) List Number of Cables	Multiply (D) × (N) Subtotal of Sum of Cable Diameters (Sd)
1.	3/C–No. 500 kcmil	2.25 inches	4	9.00 inches
2.	3/C–No. 250 kcmil	1.76 inches	3	5.28 inches
3.	3/C–No. 4/0 AWG	1.55 inches	10	15.50 inches

The sum of the diameters (Sd) of all cables (add total Sd for items 1, 2, and 3):

9.00 inches + 5.78 inches + 15.50 inches = 29.82 inches (Sd)

A cable tray with a usable width of 30 inches is required. For approximately 15% more, a 36-inch wide cable tray could be purchased that would provide for some future cable additions.

Notes:
1. Cable sizes used in this example are a random selection.
2. Cables—copper conductor with cross linked polyethylene insulation and a PVC jacket. (These cables could be ordered with or without an equipment grounding conductor.)
3. Total cable weight per foot for this installation:

   61.4 lb/ft (without equipment grounding conductors)

   69.9 lb/ft (with equipment grounding conductors)

   This load can be supported by a load symbol "B" cable tray–75 lb/ft.

Example: *NEC®* Section 318-9(a)(2)

Width selection for cable tray containing 600-volt multiconductor cables, sizes No. 3/0 AWG and smaller. Cable tray allowance fill areas are listed in Column 1 of Table 318-9.

Cable tray is obtained as follows:

METHOD 1.

The sum of the total areas for items 1, 2, 3, and 4:

3.34 sq. in. + 3.04 sq. in. + 6.02 sq. in. + 16.00 sq. in. = 28.40 sq. inches

From Table 318-9, Column 1, a 30-inch wide tray with an allowable fill area of 35 sq. inches must

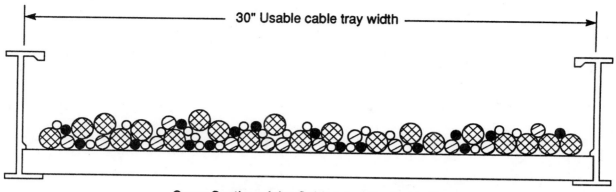

Cross Section of the Cables and the Cable Tray

(Courtesy of B-Line Systems)

be used. The 30-inch cable tray has the capacity for additional future cables (6.60 sq. inches additional allowable fill area can be used).

METHOD 2.

The sum of the total areas for items 1, 2, 3, and 4 multiplied by (6 sq. in. / 7 sq in) = cable tray width required:

3.34 sq. in. + 3.04 sq. in. + 6.02 sq. in. + 16.00 sq. in. = 28.40 sq. inches

$$\frac{28.40 \text{ sq. inches} \times 6 \text{ sq. inches}}{7 \text{ sq. inches}} = 24.34 \text{ inch cable tray width required}$$

Use a 30-inch cable tray.

Notes:
1. The cable sizes used in this example are a random selection.
2. Cables—copper conductors with cross linked polyethylene insulation and a PVC jacket. These cables could be ordered with or without an equipment grounding conductor.
3. Total cable weight per foot for this installation: 31.9 lb/ft. (Cables in this example do not contain equipment grounding conductors.) This load can be supported by a load symbol "A" cable tray– 50 lb/ft.

Example: *NEC®* Section 318-9(a)(3)

Width selection for cable tray containing 600-volt multiconductor cables, sizes No. 4/0 AWG and larger (single layer required) and No. 3/0 AWG and smaller. These two groups of cables must have dedicated areas in the cable tray.

Cable tray width is obtained as follows:

A—Width required for No. 4/0 AWG and larger multiconductor cables:

Item Number	List Cable Sizes	(D) List Cable Outside Diameter	(N) List Number of Cables	Multiply (D) × (N) Subtotal of Sum of Cable Diameters (Sd)
1.	3/C–No. 500 kcmil	2.26 inches	3	6.78 inches
2.	3/C–No. 4/0 AWG	1.55 inches	4	6.20 inches

Total cable tray width required for items 1 and 2:

6.76 inches + 6.20 inches = 12.98 inches

B—Width required for No. 3/0 AWG and smaller multiconductor cables:

Item Number	List Cable Sizes	(A) List Cable Cross Sectional Areas	(N) List Number of Cables	Multiply (A) × (N) Total of Cross Sectional Area for Each Item

Total cable tray width required for items 3, 4, and 5:

(3.20 sq. in. + 4.00 sq. in. + 3.20 sq. in.) (6 sq. in. / 7 sq. in.)1 = (10.4 sq. in.) (6 sq. in. / 7 sq. in.)1 = 8.92 inches

Actual cable tray width is A "Width" (12.98 inches) + B "Width" (8.92 inches) = 20.88 inches

A 24-inch wide cable tray is required. The 24-inch cable tray has the capacity for additional

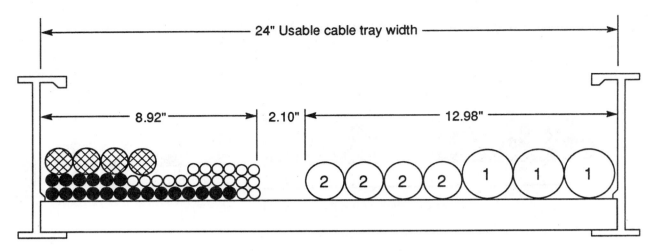

Cross Section of the Cables and the Cable Tray

(Courtesy of B-Line Systems)

future cables (3.1 inches or 3.6 sq inches allowable fill can be used).

Notes:
1. This ratio is the inside width of the cable tray in inches divided by its maximum fill area in square inches from Table 318-9, Column 1.
2. The cable sizes used in this example are a random selection.
3. Cables—copper conductor with cross linked polyethylene insulation and a PVC jacket.
4. Total cable weight per foot for this installation:

   40.2 lb/ft. (Cables in this example do not contain equipment grounding conductors.) This load can be supported by a load symbol "A" cable try–50 lb/ft.

Example: *NEC®* Section 318-9(b)

50% of the cable tray usable cross-sectional area can contain type PLTC cables.

4 inches × 6 inches × .500 = 12 square inches
allowable fill area

2/C-No. 16 AWG 300-volt shielded instrumentation cable O.D. = .224 inches.

Cross-Sectional Area = 0.04 square inches.

$$\left(\frac{12 \text{ sq. in.}}{0.04 \text{ sq. in. cable}}\right) = 300 \text{ cables can be installed in this cable tray.}$$

$$\left(\frac{300 \text{ cables}}{26 \text{ cables/row}}\right) = 11.54 \text{ rows can be installed in this cable tray.}$$

Notes:
1. The cable sizes used in this example are a random selection.
2. Cable—copper conductors with PVC insulation, aluminum/mylar shielding, and PVC jacket.

Following these general application Articles 300, 305, and 310, the remaining articles in Chapter 3 are specific articles. Article 318 covers cable trays (again see Figure 10–4). Cable trays are a support system and not a wiring method.

NOTE: As you begin to study this part of Chapter 3, refer to the definitions in Article 100 for raceway. A fine print note will alert you to the types of raceways. There is a distinct difference between raceways and cables and other wiring methods. Study that definition or refer to it as needed. A cable tray is neither a raceway nor a cable. It is merely a support system for other wiring methods.

Example: Application of Section 318-13. Ampacity of Type MV and Type MC Cables (2001 Volts or Over) in Cable Trays–(b) Single Conductor cables. These single conductor cables can be installed in a cable tray cabled together (triplexed, quadruplexed, and so on) if desired. Where the cables are installed according to the requirements of Section 318-12, the ampacity requirements are as shown in the following chart:

Sec. No.	Cable Sizes	Solid Unventilated Cable Tray Cover	Applicable Ampacity Tables *	Amp. Table Values By	Special Conditions
(1)	1/0 AWG and Larger	No Cover Allowed **	310-69 and 310-70	0.75	
(1)	1/0 AWG and Larger	Yes	310-69 and 310-70	0.70	
(2)	1/0 AWG and Larger In Single Layer	No Cover Allowed **	310-69 and 310-70	1.00	Maintained Spacing Of One Cable Diameter
(2)	Single Conductor in Triangle Config. 1/0 AWG and Larger	No Cover Allowed **	310-71 and 310-72	1.00	Maintained Spacing Of 2.15 × One Conductor O.D.

 * The ambient ampacity correction factors must be used.

** At a specific position where it is determined that the tray cables require mechanical protection, a single cable tray cover of six feet or less in length can be installed.

As needed, refer to the articles from 320 through 365 for the specific types of raceways and cables. In addition to the references for raceways in Chapter 3, you will find references related to Chapter 9 tables and examples, such as the reference found in Section 345-7, the numbers of conductors in conduit.

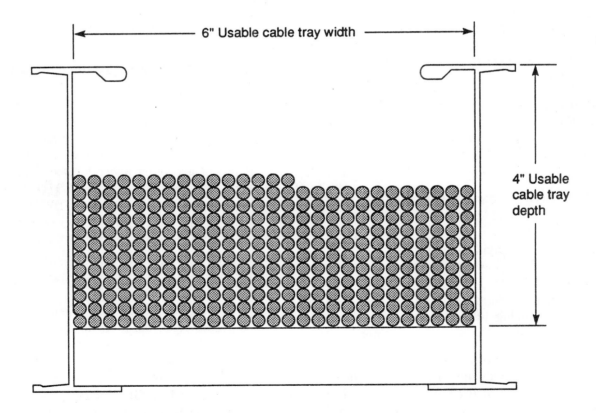

Cross Section of the Cables and the Cable Tray

(Courtesy of B-Line Systems)

Example: A typical 480/277-volt, 400-ampere feeder is run from a distribution unit substation under a concrete slab exposed above a suspended ceiling turning down into a panelboard. Based on the length (200 feet) of the run and the calculated load, the installer chooses to pull three 600 kcmil THWN ungrounded circuit conductors and a full-sized 600 kcmil THWN neutral (grounded conductor). If the installer chooses a nonmetallic raceway for a portion of this run, he recognizes that he will also have to add an equipment grounding conductor sized in accordance with Table 250-122 or a No. 3 THWN copper. However, when he applied the tables in Chapter 9, he found he would need the following sizes of conduit or tubing to make this installation.

Solution: Table 5. 600 kcmil THWN = 0.8676 sq. in.

4×0.8676	= 3.47 sq. in.

Table 4. EMT Electrical Metallic Tubing 40% fill as per Table 1	= **3 inch**
IMC Intermediate Metal Conduit 40% fill as per Table 1	= $3\frac{1}{2}$ inch
GRC Galvanized Rigid Conduit 40% fill as per Table 1	= $3\frac{1}{2}$ inch

Solution: Table 5. 600 kcmil THWN = 0.8676 sq. in.

4×0.8676	= 3.47 sq. in.
Plus an equipment grounding conductor No. 3 THWN	= $\underline{0.0973}$ sq. in.
	3.5673 sq. in.

Table 4. ENT Electrical Nonmetallic Tubing Max. size 2 inch	= N/A
Sch. 40 Rigid Nonmetallic Conduit 40% fill as per Table 1	= $3\frac{1}{2}$ inch
Sch. 80 Rigid Nonmetallic Conduit 40% fill as per Table 1	= **4 inch**

Using these same calculations with 500 kcmil for the ungrounded and grounded conductors would require a 3-inch conduit for EMT, IMC, GRC, and PVC Schedule 40, and a $3\frac{1}{2}$ inch for PVC Schedule 80. As one can see, calculations will be necessary to determine the most efficient wiring method.

When sizing the number of conductors in a raceway, conduit, tubing, and so on, it is necessary to refer to the permitted percent fills specified in the tables in Chapter 9, Table 1 for percentage fill and Table 4 for the conduit dimensions. Similar references are found in Section 346-7 for rigid metal conduit, Section 347-11 for rigid nonmetallic conduit, Section 348-8 for electrical metallic conduit, and Section 350-12 for flexible metallic conduit. Other similar references are found for other raceway types. The number of conductors allowed in a given raceway must be calculated in square inches, based on the insulation and wire size listed in Tables 5 and 5-a of Chapter 9. For bare conductors, Chapter 8 dimensions are used. Where conductors of all the same size are installed in a single raceway, Table 3 applies generally. The following example will help you learn to apply Chapter 9.

Conduit & Tubing Diameters will make a difference in the 1999 NEC®
½ IMC intermediate metal conduit internal diameter .0660 inch

Allowable 40% fill would permit three No. 8 Type THWN Conductors
½ Schedule 80 rigid nonmetallic conduit internal diameter 0.526 inch

Allowable 40% fill would permit one No. 8 Type THWN conductor

---

Example: What size PVC rigid nonmetallic conduit is required for a multiple branch-circuit supplying a cooling tower located adjacent to a hospital, containing three No. 6 THWN stranded copper conductors, three No. 4 THWN copper conductors, four No. 1/0 THW copper conductors, and one No. 2 bare copper conductor?

Step 1. Rigid nonmetallic conduit—Article 347. Number of conductors—Section 347-11, Reference Table 1, Chapter 9.

Step 2. Chapter 9, Notes 1 and 2, and Table 1 and Notes. Read the notes to the table carefully because they are a part of the table and, where applicable, they are mandatory.

Step 3. Chapter 9, Table 1—Over two conductors permits a 40% fill of the raceway.

Step 4. Table 4 dimensions and percent are of conduit and tubing is in square inches. Therefore, select square inches dimensions from Tables 5 and 8 for the correct calculations.

Step 5.   Table 5—
3 No. 6 THWN = .00507 × 3 = .1521 sq. inch
3 No. 4 THWN = .0824 × 3 = .2472 sq. inch
4 No. 1/0 THW = .2223 × 4 = .8892 sq. inch
Table 8-1—
No. 2 BARE   = .067 × 1   = .067 sq. inch

Total   1.3555 sq. inches

Table 4—Not lead covered over two conductors in a raceway 40% 2-inch rigid PVC Schedule 40 raceway is 1.316 sq. inches. A 2 ½ raceway is 1.878. Therefore, two inches is not large enough and it is necessary to install 2 ½ inch rigid nonmetallic conduit for this installation.

NOTE: Two-inch intermediate metal conduit (IMC) (Article 345) would permit a 40% fill of 1.452 square inches. Therefore, 2-inch IMC could be used.

Article 370 covers outlet boxes, pull boxes, junction boxes, conduit bodies, and fittings. A conduit body is what is normally termed a Tee, LB, and so forth, or as a condulet. Use Article 370 carefully, and as you determine the number of conductors for a given box, the sizing requirements are found in Section 370-16.

NOTE: Most boxes with the integral type clamps, such as nonmetallic outlet boxes, are clamps and a deduction must be added. If uncertain, check with the manufacturer. In a test, normally, the test material will tell you whether there are clamps to be calculated. Study Section 370-16(a) and (b) and the tables carefully.

Section 370-16(a)(2)

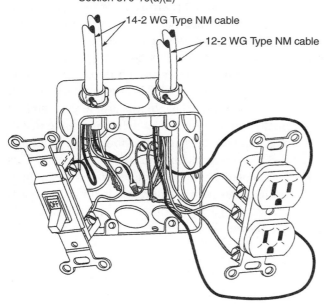

14-2 WG Type NM cable
12-2 WG Type NM cable
OFF

Example: What size outlet box with integral clamps is required to accommodate a 15-ampere snap switch to control lighting and a 20-ampere duplex receptacle? The switch has a 2-wire w/GRD No. 14 nonmetallic sheathed cable feeding it with another No. 14 NM 2-wire w/GRD No. 12 feeding the light fixture. The receptacle is fed with a 2-wire w/GRD No. 12 nonmetallic sheath cable and a 2-wire w/GRD nonmetallic sheath cable goes on to feed additional duplex receptacles in the same area.

Solution:

Section 370-16(a)(1) & (2). The minimum size of box in cubic inches is based on the following calculations.

The required deduction for each type fitting (i.e., clamps, hickeys, etc.) and all equipment grounding conductors require one deduction for each based on the largest conductor entering the box, Section 370-16(a)(1), (2), and Table 370-16(b).

In this example, a No. 12 is the largest. Therefore, a 2.25 cubic inch deduction is required for each.

Equipment grounding conductors = 2.25
Cable clamps = No. 12.   2.25 × 1 = 2.25

Each device requires two deductions
each, based on size conductor
connected to the device.

Receptacle = No. 12.   2 × 2.25   = 4.50
Switch   = No. 14.   2 × 2.00   = 4.00

Conductors entering the box,
one deduction each:

No. 12.   2.25 × 4   = 9.00
No. 14.   2.00 × 4   = 8.00

The minimum size box permitted would be a 30 cubic inch box.

In addition, sizing pull boxes and junction boxes will be in most examinations. These requirements are found in 370-28.

Example: Four 3-inch rigid steel conduits or EMT, straight pull. Three 2-inch rigid steel conduits or EMT, angle pull.

Recommended minimum spacing for 3-inch conduit, 4¾ inches; for 2-inch conduit, 3⅜ inches; and for three 2-inch conduits, 4 inches.

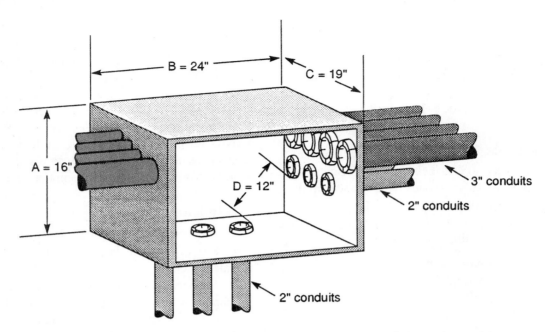

Pull and junction boxes. Minimum size as required in *NEC®* Section 370-28.

For depth of box (A)

Three 2-inch conduits, angle pull:

$$A = (6 \times 2) + 2 + 2 \qquad = 16 \text{ inches}$$

Spacing for two rows of conduit:

Edge of box to center of 3-inch conduit	$= 2\frac{3}{8}$ inches
Center to center, 3- to 2-inch conduit	$= 4\frac{1}{2}$
Minimum distance, D	$= \underline{8\frac{1}{2}}$
Total	$14\frac{7}{8}$ inches

Required depth: Use 16 inches

For length of box (B)

Four 3-inch conduits, straight pull: $B = 8 \times 3$	$= 24$ inches
2-inch conduit, angle pull: $B = A$	$= 16$ inches

Required length: Use 24 inches

For width of box (C)

Recommended spacing for 3-inch conduits	$= 4\frac{3}{4}$ inches
$C = 4 \times 4\frac{3}{4}$	$= 19$ inches

Required width: Use 19 inches

Angle pull distance (D)

$$D = 6 \times 2 \qquad = 12 \text{ inches}$$

Required distance: Use 12 inches

This section also should be studied, and you must know where to find it quickly so you can determine the size, outlet, or junction box as required. Article 373 covers cabinet, cutout boxes, and meter socket enclosures. Most panelboards, switches, disconnecting switches, conductors, and so on, are enclosed in a cabinet or cutout box. When determining the requirements for cabinet and cutout boxes, refer to Article 373. Article 374 covers auxiliary gutters. Auxiliary gutters are not raceways or cables. An auxiliary gutter is an extension used to supplement wiring spaces in meter enclosures, motor control centers, distribution centers, switchboards, and similar wiring systems, and can enclose conductors or bus bars but is not permitted for switches, overcurrent devices, appliances, or other similar equipment. Although similar in appearance to a wireway, an auxiliary gutter has a different application and use. Be aware of Article 374 for auxiliary gutters. Article 380 covers switches. In this article you can find requirements for all types of switches and some installation requirements, such as accessibility and grouping. Article 384 covers switchboards and panelboards. This article undoubtedly will be included in most examinations and includes many requirements for panelboards and switches. Of particular importance is Section 384-4, an installation requirement. You are urged to mark in your Code book in Section 110-16 to also refer to Section 384-4 for the dedicating working space above and below equipment. This section also will be in most examinations. You need not try to memorize this section, because it is specific requirements for this equipment, and your proficiency in using this as a reference document will adequately guide you through these wiring method articles. Be familiar with the arti- cle names as listed in the Table of Contents and it will assist you as you use this book for reference or for a tool in the field, and will increase your proficiency in preparing for an examination.

CHAPTER 10 QUESTION REVIEW

> **Each lesson is designed purposely to require the student to apply the entire *NEC*® text, not specific chapters or articles. It has been found that when studying for a timed, open-book examination the student must gain proficiency in the Table of Contents, the Index, and the ability to move quickly from cover to cover to find the correct answers to each question in a timely fashion.**

1. When installing Type SER service entrance cable in a building from the service panel to a dryer outlet, what are the support requirements for this cable?

   Answer: _____

   _____

   _____

   _____

   Reference: _____

2. When installing conduit bodies that are legibly marked with their cubic inch capacity, how can one determine the number of conductors and splices and taps permitted?

   Answer: _____

   _____

   _____

   _____

   Reference: _____

3. Is ⅜ liquidtight flexible metal conduit permitted for enclosing motor leads in accordance with Section 430-145(b)?

   Answer: _____

   _____

   _____

   _____

   Reference: _____

4. Structural bar joists are often spaced at up to 5-foot intervals. Support is difficult to achieve when installing metal raceways on these bar joists. Is it permitted to mount an outlet box on a bar joist and extend intermediate metal conduit (IMC) 5 feet to the first support?

    Answer: _____

    _____

    _____

    _____

    Reference:_____

5. When installing electrical nonmetallic tubing (ENT), is it permissible to install this material through metal studs without securing the ENT every 3 feet as required by the Code?

    Answer: _____

    _____

    _____

    _____

    Reference:_____

6. What are the support requirements and methods of securing Type NM cable?

    Answer: _____

    _____

    _____

    _____

    Reference:_____

7. Can nonmetallic wireways be installed in hazardous Class 1 or Class 2 locations?

    Answer: _____

    _____

    _____

    _____

    Reference:_____

8. What wiring methods are permitted for installing a branch-circuit routed through the bar joists of a metal building for a 4,160-volt, 3-phase motor?

Answer: _____

_____

_____

_____

Reference: _____

9. You have a contract to install a branch-circuit under a bank president's desk to serve his or her calculator. The bank has asked that you not drill any holes in the walls. Can you lay the conductors under the carpet? If so, what type of wiring method should you use?

Answer: _____

_____

_____

_____

Reference: _____

10. When installing wireway metallic or nonmetallic, either type, how do you derate the conductors when exceeding the number permitted?

Answer: _____

_____

_____

_____

Reference: _____

11. When installing Type AC cable on wood studs, routing them parallel to the framing members, what distance clearance must be maintained where the Type AC cable is likely to be penetrated by screws or nails?

Answer: _____

_____

_____

_____

Reference: _____

12. How do the installation requirements for intermediate metal conduit (IMC) and that of rigid metal conduit (RMC) differ?

Answer: _____

_____

_____

_____

Reference: _____

13. What is the minimum size conductor that can be run in parallel, generally? (Do not consider exceptions.)

Answer: _____

_____

_____

_____

Reference: _____

14. Where installing electrical nonmetallic tubing for the branch-circuit conductors in a five-story building, does the material from which the walls, floors, and ceilings are constructed need to be considered?

Answer: _____

_____

_____

_____

Reference: _____

15. Christmas tree lights have been installed on the local courthouse. The mayor suggested that we leave them up throughout the year. Is this permitted? How long are they permitted to remain installed?

Answer: _____

_____

_____

_____

Reference: _____

16. When installing a run of conduit, it is necessary to install a piece of liquidtight flexible metal conduit 6 feet long through a plenum. Is this permitted by the Code? How much liquidtight flexible metal conduit can you install in a plenum?

Answer: _____

_____

_____

_____

Reference: _____

17. Where making an installation of rigid nonmetallic conduit, Schedule 80, along the outside of a building to feed outdoor floodlights and studying the Code, you find that in Section 300-7(b) expansion joints must be provided where necessary. How much expansion or contraction must rigid nonmetallic conduit have before an expansion fitting is required?

Answer: _____

_____

_____

_____

Reference: _____

18. A column panel is a narrow panel that is installed inside or flush with an I-beam. It is primarily used in warehouses and areas where the panel is subjected to moving traffic and warehousing storage, such as forklifts and palletizing. When installing a column-type panel, an auxiliary gutter is installed above the panel, and the neutral connections are made up at the top of the auxiliary gutter and not brought down into the panelboard. Is this an acceptable wiring method?

Answer: _____

_____

_____

_____

Reference: _____

19. When wiring a single-family dwelling to save space, a 2-inch rigid conduit is installed from the top of the panelboard into the attic space and pulled the nonmetallic sheath cables within that 2-inch conduit. Is this a permitted wiring method?

Answer: _____

_____

_____

_____

Reference: _____

20. A small lake cabin is to be wired. This cabin will contain wood heating, no air conditioning, and only six single-pole branch-circuits for lighting and receptacle outlets. Using only six switches in this panel, are you exempt from using a main circuit breaker or a single disconnecting switch as required in Chapter 2 for services?

Answer: _____

_____

_____

_____

Reference: _____

Chapter Eleven

GENERAL USE UTILIZATION EQUIPMENT

The equipment covered in Chapter 4 is generally found in most facilities—residential, commercial, and industrial. Chapter 4 begins with two articles about wire. One might think that these articles are misplaced and should be found in Chapter 3. However, flexible cords and cables and fixture wires are usually associated with general purpose equipment, such as flexible cords for appliances, motors, and fixture wiring is usually associated with lighting fixtures.

---

1. The maximum wattage permitted for a medium-base incandescent lamp (standard light bulb) is _____ watts.

2. The maximum wattage permitted for a mogul-base incandescent lamp is _____ watts.

3. A special base or other means must be used for incandescent lamps of _____ watts and larger.

Answers: (1) 300, (2) 1,500, (3) 1,501.
Reference: Section 410-53

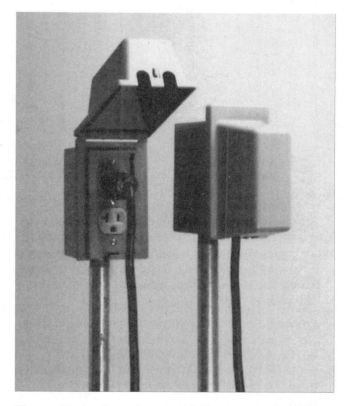

Figure 11–1 Compliance with Section 410-57. *(Courtesy of TayMac Corporation)*

Chapter 4 covers all types, lighting fixtures, appliances, fixed electric space heating equipment, motors, transformers, air-conditioning and refrigeration equipment, and several special use types of general equipment, such as phase converters often used in rural farm-type installations, and capacitors now generally found in many types of equipment that improve the efficiency of many services as a power factor correction. Generators are often used to supplement or provide emergency power or standby power. As you study Chapter 4 of the *NEC*®, be aware that the requirements in 410 cover not only lighting fixtures but also cover receptacles, cord connectors, and attachment caps. (See Figure 11–1 and Figure 11–2.) However, Article 410 does not cover signs. (See Article 600 for electric signs and outline lighting.) Article 422, which covers the appliances, generally has specific requirements for branch-circuits, installation of appliances, and the control and protection of appliances. Space heating, motors, air-conditioning and refrigeration equipment articles also have general provisions and installation provisions within those articles.

Section 422-7 states that central heating equipment must be supplied by an individual branch-circuit. What is an individual branch-circuit?

Answer: An individual branch-circuit is defined in Article 100 as a branch-circuit that supplies only one utilization equipment.

WET LOCATIONS

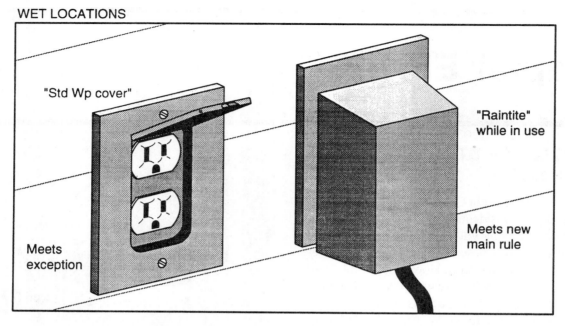

Figure 11–2   *(Courtesy of TayMac Corporation)*

Would it be permissible to connect a waste (garbage) disposal and a built-in dish washer with flexible cords? Both cords would be a minimum of 42 inches.

Answer: Section 422-16(b)(2) would permit the built-in dishwasher to be connected with a cord 3 feet to 4 feet long. Section 422-16(b)(2), however, would prohibit the 42-inch cord on the waste disposal. This section limits the cord for disposals to 18 to 36 inches.

NOTE: Portable dishwashers are supplied by the manufacturer with a cord and plug attached. This cord is evaluated as a part of the listing and length is not limited by Section 422-16.

A new article, Article 490, has been added for equipment over 600 volts. It covers the general requirements for equipment operating at more than 600 volts and defines "high voltage" as more than 600 volts. These requirements were formerly found in Article 710, which has been deleted.

Read Section 430-6 Carefully

The motor nameplate current is not to be used. The values in Tables 147 through 150 are to be used instead.

Example: Motor branch-circuit conductors ampacities are as per Article 310. However, Section 240-3(d) limiting No. 14 to a 15-ampere overcurrent device, No. 12 to a 20-ampere device, and No. 10 to a 30-ampere device do **not** apply. Section 430-6 only references the table and not Section 240-3(d).

Example: In a steel mill, a number of small 3-horsepower, 3-phase, 480-volt motors are supplied by a 400-ampere feeder. What is the minimum size 10-foot tap conductor permitted to be used to feed each motor?

Step 1. Section 240-3(e), which permits tap conductors to be protected against overcurrent in accordance with Sections 240-19(c), 240-21, 364-11, 364-12, and 430-53(d).

Step 2. Section 240-21(f) references Sections 430-28 and 430-53(d) for motor taps.

Step 3. Section 430-28.   400 amperes ÷ 1000% = 40 amperes.

Step 4. Table 310-16 answer No. 8 THWN

NOTE: Some would permit a No. 10 to be used in this calculation as taken from the 90° column because a 3-horsepower motor (Table 430-150) would only draw 4.8 amperes, and the temperature would not be exceeded.

CHAPTER 11 QUESTION REVIEW

> **Each lesson is designed purposely to require the student to apply the entire *NEC®* text, not specific chapters or articles. It has been found that when studying for a timed, open-book examination the student must gain proficiency in the Table of Contents, the Index, and the ability to move quickly from cover to cover to find the correct answers to each question in a timely fashion.**

THREE-PHASE MOTOR AS FOLLOWS:
Type: squirrel cage, high reactance
Voltage: 460 volts, 3-phase
HP: 30
Current: nameplate is the same as *NEC®* current Tables 430-147-150
Duty: continuous
Start: high reactance
Code letter: none
Service factor: none

Refer to the 3-phase motor above for Questions 1–3.

1. What *NEC®* section requires that the Tables 430-147-150 be used instead of actual current rating marked on motor nameplate?

Answer: _____

_____

_____

_____

Reference: _____

2. What size inverse time circuit breaker is needed to provide the maximum branch-circuit, short-circuit, ground-fault protection?

Answer: _____

_____

_____

_____

Reference: _____

3. A dual element (time delay) fuse can be sized a maximum of _____ % when used as branch-circuit protection.

Answer: _____

_____

_____

_____

Reference: _____

4. Where branch-circuit conductors are tapped to No. 12 with a 20-ampere circuit rating as permitted in the *NEC®*, the tap cannot be smaller than size _____.

    Answer: _____

    _____

    _____

    _____

    Reference: _____

In Questions 5 through 8 calculate the ampere rating of the fuse size needed for protection of the following copper conductors where no motors are connected:

5. No. 12 THHN _____ ampere rated fuse

    Answer: _____

    _____

    _____

    Reference: _____

6. No. 8 TW _____ ampere rated fuse

    Answer: _____

    _____

    _____

    Reference: _____

7. No. 3 THW _____ ampere rated fuse

    Answer: _____

    _____

    _____

    Reference: _____

8. No. 3/0 THWN _____ ampere rated fuse

    Answer: _____

    _____

    _____

    Reference: _____

9. For a transformer 600 volts or less that has no secondary overcurrent protection, the primary overcurrent protection shall not exceed _____% of the primary current.

   Answer: _____

   _____

   _____

   _____

   Reference: _____

10. The maximum branch-circuit and ground-fault protection using time-delay fuses for typical motors are sized at _____% of the motor's full-load current.

    Answer: _____

    _____

    _____

    _____

    Reference: _____

11. Overload protection (heaters, relays, time-delay fuses, thermal overloads) for typical motors having a service factor of 1.15 is generally sized at _____ % of the motor's full-load current.

    Answer: _____

    _____

    _____

    _____

    Reference: _____

12. A motor controller that is installed with the expectation of its being submerged in water occasionally for short periods shall be installed in a rated enclosure type No. _____.

    Answer: _____

    _____

    _____

    _____

    Reference: _____

13. Where recessed high-intensity discharge fixtures are installed _____ and operated by remote ballasts, both the fixture and the ballast require thermal protection.

    Answer: _____

    _____

    _____

    _____

    Reference: _____

14. Any pipe or duct system foreign to the electrical installation must not enter a transformer vault. The _____ piping is not considered foreign to the vault.

    Answer: _____

    _____

    _____

    _____

    Reference: _____

15. A 10 kVA dry-type transformer rated at 480 volts can be installed on a building column and shall not be required to be _____.

    Answer: _____

    _____

    _____

    _____

    Reference: _____

16. A capacitor is located indoors. It must be enclosed in a vault if it contains more than a minimum of _____ gallon(s) of flammable liquid.

    Answer: _____

    _____

    _____

    _____

    Reference: _____

17. A single-phase hermetic refrigerant motor-compressor has a rated load current of 24 amps and a branch-circuit selection current of 30 amps. The branch-circuit conductors are copper, type TW. They operate at 80°F, and they are the only conductors in the conduit to this compressor. The smallest possible branch-circuit conductors must be least size _____ AWG.

    Answer: _____

    _____

    _____

    _____

    Reference: _____

18. A dry-type transformer is to be installed indoors. If rated more than ____ kilovolt-amps, the transformer must be installed in a fire-resistant transformer room.

Answer: _____

_____

_____

_____

Reference:_____

19. An electric resistance heater is rated for 2,400 watts at 240 volts. What power is consumed when the heater is operated at 120 volts?

Answer: _____

_____

_____

_____

Reference:_____

20. A 230-volt single-phase circuit has 10 kilowatts of load and 50 amps of current. The power factor is ____ %.

Answer: _____

_____

_____

_____

Reference:_____

Chapter Twelve

SPECIAL EQUIPMENT AND OCCUPANCIES

As you begin to study Chapter 12, remember that Chapter 11 covered general equipment, equipment that is found in most every common installation, equipment that is for general use, such as light fixtures, receptacles, motors, transformers, and so forth. In this chapter, we note that we are studying special occupancies, special equipment, and special conditions. These rules apply to these items and are not generally found in all buildings. They are special. Study carefully Article 90, Chapter 1, and Article 300 of the text. Article 90-3 outlines the use of the *NEC*® and states that Chapters 1 through 4 apply generally and Chapters 5, 6, and 7 amend, supplement, or modify Chapters 1 through 4.

> **Caution:** *NEC*® 90-3, Chapter 5 only modifies and augments Chapters 1 through 4. Corrosion, Capacity, Suitability, and Service are not considered. Therefore, it is the responsibility of the designer to take these factors into account along with the hazardous location requirements covered in Chapter 5.

Special occupancies include such things as hazardous areas, bulk storage plants, service stations, paint spray booths, commercial garages, and many more hazardous occupancies. They also include occupancies, such as hospitals, mobile homes, and agriculture buildings. On the other hand, Chapter 6 deals with special equipment, such as signs, not found in every building but still common equipment. Swimming pools, welding equipment, electric welders, elevators, dumbwaiters, office furnishings, and many more types are covered.

Chapter 7 deals with special conditions. These are conditions that are special and not general conditions, such as emergency systems, and legally and optional standby systems, over-600-volt systems, optical fiber cables, and raceways. Finally, in Chapter 7 is Article 780, which deals with the intelligent buildings, such as Smart House, the closed loop systems that offer a totally different concept to the safe use of electricity as we have known it in the past. Study the questions

carefully. Remember as you study these chapters, they modify, supplement, or amend Chapters 1 through 4 of the *NEC*®.

Examination Study

Questions related to Chapter 6, Special Equipment, can generally be answered quickly because these articles are usually short and cover very specific requirements. The most important study advice is to learn the names of these articles for quick reference.

- Article 600. Signs and Outline Lighting
- Article 605. Office Furnishings
- Article 610. Cranes and Hoists
- Article 620. Elevators, Dumbwaiters, Escalators, Moving Walks, Wheelchair Lifts, and Stairway Chairlifts
- Article 625. **New to the 1996 *NEC*®.** Electric Vehicle Charging System Equipment. This new article will place installation requirements for electrical powered automobiles in the Code. This article does not cover other charging systems, such as golf carts or forklifts. It does cover individual residential-type chargers and the proposed commercial quick-charging stations that may someday appear at rest stops along the interstate highway system. It is said that someday soon, you will be able to stop for coffee and get your electric car charged in a few minutes.
- Article 630. Electric Welders
- Article 640. Sound Recording and Similar Equipment
- Article 645. **New Title "Information Technology Equipment."** Formerly "Computer/Data Processing Equipment."
- Article 650. Pipe Organs
- Article 660. X-ray Equipment
- Article 665. Induction and Dielectric Heating Equipment
- Article 668. Electrolytic Cells

- Article 669. Electroplating
- Article 670. Industrial Machinery
- Article 675. Electrical-Driven or Controlled Irrigation Machinery
- Article 680. Swimming Pools, Fountains, and Similar Equipment

- Article 685. Integrated Electrical Systems
- Article 690. Solar Photovoltaic Systems
- Article 695. Fire Pumps. **New to the 1996 NEC®.** This article has added installation requirements into the Code that were formerly found in NFPA 20.

CHAPTER 12 QUESTION REVIEW

Each lesson is designed purposely to require the student to apply the entire *NEC®* **text, not specific chapters or articles. It has been found that when studying for a timed, open-book examination the student must gain proficiency in the Table of Contents, the Index, and the ability to move quickly from cover to cover to find the correct answers to each question in a timely fashion.**

1. Name three types of optical fiber cable.

   Answer: _____

   _____

   _____

   _____

   Reference: _____

2. Fixed wiring over a Class I location in a commercial garage cannot be enclosed in nonmetallic sheathed cables. (True or False)

   Answer: _____

   _____

   _____

   _____

   Reference: _____

3. Are seals required for outdoor propane-dispensing unit located 50 feet from the office where the branch-circuit supplying the unit originates? If the answer is yes, where are they to be placed?

   Answer: _____

   _____

   _____

   _____

   Reference: _____

4. Emergency electrical systems are those systems legally required and classed as emergency by Article 700 *NEC®*? (True or False)

Answer: _____

_____

_____

_____

Reference:_____

5. In a commercial garage, Class _____ Division _____ wiring methods must be met for an electrical outlet installed 12 inches above the floor if there is no mechanical ventilation?

Answer: _____

_____

_____

_____

Reference:_____

6. According to the *National Electrical Code®*, transfer equipment for emergency systems shall be designed and installed to prevent what?

Answer: _____

_____

_____

_____

Reference:_____

7. The sealing compound in a completed conduit seal for a Class I location must be at least _____ inch thick.

Answer: _____

_____

_____

_____

Reference:_____

8. Bare conductors are field-connected to fixed terminals of different phases on an outdoor, 13.8 kV circuit. According to the *National Electrical Code®*, the air separation between these conductors shall be a minimum of _____ inches.

Answer: _____

_____

_____

_____

Reference:_____

9. In a hospital's general care areas, the number of receptacles required to be in a patient-bed location is a minimum of _____.

Answer: _____

_____

_____

_____

Reference: _____

10. A 120-volt single-phase sign panel supplies only the sign circuits. It is on 24 hours a day. It supplies three ⅓ horsepower motors, 36-0.80 amp ballasts. Each ungrounded conductor in the subfeeder to the sign panel must be at least No. _____ AWG copper THWN.

Answer: _____

_____

_____

_____

Reference: _____

11. Portable gas tube signs for interior use can be connected with a supply cord having a maximum length of _____ meters. (_____ feet)

Answer: _____

_____

_____

_____

Reference: _____

Answer Questions 12 through 15 about Type S fuses True or False.

12. A 15-ampere Type S fuse will fit into a 20-ampere Type S adapter.

Answer: _____

_____

_____

_____

Reference: _____

13. A 20-ampere Type S fuse will fit into a 20-ampere Type S adapter.

    Answer: _____

    _____

    _____

    _____

    Reference: _____

14. A 25-ampere Type S fuse will fit into a 30-ampere Type S adapter.

    Answer: _____

    _____

    _____

    _____

    Reference: _____

15. A $6\frac{1}{4}$ Type S adapter will accept a 3 amp, 4, $4\frac{1}{2}$, 5, $5\frac{6}{10}$, and $6\frac{1}{4}$ ampere Type S fuse.

    Answer: _____

    _____

    _____

    _____

    Reference: _____

16. The *NEC*® requires that the available fault current be marked when series rated systems are installed. (a) Who is required to install that marking? (b) What must the marking state? (c) Who is responsible for providing this information?

    Answers: _____

    _____

    _____

    _____

    Reference: _____

17. A $1\frac{1}{2}$ horsepower single-phase motor, which has an efficiency of 80%, operates at 230 volts, and has an input current of _____ amps. (1 hp = 746 watts.)

    Answer: _____

    _____

    _____

    _____

    Reference: _____

18. A 30-horsepower wound-rotor induction motor with no code letter is to be installed with 460-volt, 3-phase, alternating current. Disregarding all the exceptions, the nontime delay fuse for short-circuit protection of the motor branch-circuit must be rated at a maximum _____ amps.

    Answer: _____

    _____

    _____

    _____

    Reference: _____

19. The following 480-volt, 3-phase, 3-wire, intermittent use equipment is in a commercial kitchen: two 5,000-watt water heaters, four 3,000-watt fryers, and two 6,000-watt ovens. Each ungrounded conductor in the feeder circuit for this kitchen equipment must be sized to carry a minimum computed load of _____ amps.

    Answer: _____

    _____

    _____

    _____

    Reference: _____

20. A 240/480 3-phase power panelboard supplies only one 15,000 VA, 480-volt, 3-phase balanced resistive load. Each ungrounded conductor in the subfeeder to this power panel has a total net computed load of _____ amps.

    Answer: _____

    _____

    _____

    _____

    Reference: _____

Chapter Thirteen

SPECIAL CONDITIONS AND COMMUNICATION SYSTEMS

The first three articles in Chapter 7 cover emergency systems (Article 700), legally required standby systems (Article 701), and optional standby systems and other special systems (Article 702). The determination for which type system is installed in a building or facility is made by others, i.e., Fire Marshal, Building Official, etc. The *NEC®* does not make these requirements. It does not require exit lights in buildings. It does not require smoke detectors. However, when these items are required to be installed, the *NEC®* governs. The installation requirements must conform to these articles as appropriate.

Article 710, which covered the general requirements for systems and circuits over 600 volts nominal, has been deleted in the 1999 *NEC®*. The contents were relocated to the appropriate articles. This was done to recognize that installations of systems and circuits over 600 volts are common and are correctly placed in Chapters 1 through 4 of the *NEC®*, which apply generally.

- A new Article 490 titled "Equipment, over 600 volts nominal" has been developed and added. It will apply to the general requirements for equipment over 600 volts.
- The wiring method requirements are now located in Article 300 Part B Sections 300-31 through 300-50.
- Overcurrent requirements have been relocated to Article 240 Part I Sections 240-100 and 240-101.
- Work space and accessibility requirements have been relocated into Article 110 Part C and have been added to the existing working space requirements that were already located in this section of the *NEC®*.
- Tunnel installations over 600 volts nominal have been relocated to a new Part D to Article 110.
- The grounding requirements were deleted and Article 250 now applies as appropriate.

Article 720 covers the few circuits and equipment operating at less than 50 volts that are not already covered in the Code by Articles 411, 551, 650, 669, 690, 725, and 760.

Article 725 covers power-limited circuits of Class 1, 2, and 3. It is very important that one understand the parameters of each class of wiring system. The power is limited by the power source, generally a transformer.

Class 1
Source rated output no more than 30 volts and 1000 volt amperes

Class 2 and 3
Source rated output no more than
   (1) A listed Class 2 or Class 3 transformer.
   (2) A listed Class 2 or Class 3 power supply.
   (3) Other listed equipment marked to identify the Class 2 or Class 3 power.

   (FPN): Examples of other listed equipment are a circuit card listed for use as a Class 2 or Class 3 power source, where used as part of a listed assembly; a current-limiting impedance, listed for the purpose or part of a listed product, used in conjunction with a nonpower-limited transformer or a stored energy source, e.g., storage battery, to limit the output current; a thermocouple.

   Exception: Thermocouples shall not require listing as a Class 2 power source.

   (4) Listed information technology (computer) equipment limited power circuits.

   (FPN): One way to determine applicable requirements for listing of information technology (computer) equipment is to refer to UL 1950, "Safety of Information Technology Equipment, Including Electrical Business Equipment." Typically, such circuits are used to interconnect information technology equipment for the purpose of exchanging information (data).

   (5) A dry cell battery shall be considered an inherently limited Class 2 power source, provided the voltage is 30 volts or less and the capacity is equal to or less than that available from series connected No. 6 carbon zinc cells.

(FPN): Tables 11(a) and 11(b) in Chapter 9 provide the requirements for listed Class 2 and Class 3 power sources.

For example, Class 2 conductors run through a plenum must be run as Type CL2P, MPP, CMP, FPLP, or with approriate building wire run in rigid, IMC, EMT Type MI cable, or specific types of solid sheathed MC cable. (See Section 300-22.)

PLFA conductors run through a plenum must be run as Type FPLP, MPP, CMP, CL3P, or with appropriate building wire run in rigid, IMC, EMT Type MI cable, or specific types of solid sheathed MC cable. (See Section 300-22.)

Reprinted with permission from NFPA 70-1999.

Section 770-6.

Raceways for Optical Fiber Cables. *The raceway shall be of a type permitted in Chapter 3 and installed in accordance with Chapter 3.*

Exception: Listed nonmetallic optical fiber raceway identified as general purpose, riser, or plenum optical fiber raceway in accordance with Section 770-51 and installed in accordance with Sections 331-7 through 331-14, where the requirements applicable to electrical nonmetallic tubing shall apply. Unlisted underground or outside plant construction plastic inner duct shall be terminated at the point of entrance.

FPN: For information on listing requirements for optical fiber raceways, see Standard for Optical Fiber Raceways, UL 2024.

Where optical fiber cables are installed within the raceway without current-carrying conductors, the raceway fill tables of Chapter 3 and Chapter 9 shall not apply.

Where nonconductive optical fiber cables are installed with electric conductors in a raceway, the raceway fill tables of Chapter 3 and Chapter 9 shall apply.

Reprinted with permission from NFPA 70-1999.

Article 780 titled "Closed-Loop and Programmed Power Distribution," often referred to as "Smart House," applies to premise power distribution systems jointly controlled by a signaling between the energy-controlling equipment and utilization equipment. All equipment, including the conductors, are required to be listed and identified. It will apply only for signaling circuits not exceeding 24 volts; the current required shall not exceed 1 ampere where protected by an overcurrent device or an inherently limited power source. The article also contains specific safety requirements for this new system.

Listed hybrid cable is permitted that may contain power, communications, and signaling conductors under a common jacket. The jacket must be applied so as to separate the power conductors from the communications and signaling conductors. The individual conductors of a hybrid cable shall conform to the Code provisions applicable to their current, voltage, and insulation rating. The signaling conductors shall not be smaller than No. 24 copper. The power, communications, and signaling conductors of listed hybrid cable are permitted to occupy the same cabinet, panel, or outlet box (or similar enclosure housing the electrical terminations of electric light or power circuits) only if connectors specifically listed for hybrid cable are employed.

Chapter 8 is a stand-alone chapter in that Section 90-3 states

---

Chapter 8 covers communications systems and is independent of the other chapters except where they are specifically referenced therein.

This chapter is divided into three articles, Article 800 for:

---

telephone, telegraph (except radio), outside wiring for fire alarm and burglar alarm, and similar central station systems; and telephone systems not connected to a central station system but using similar types of equipment, methods of installation, and maintenance.

(FPN No. 1): For further information for fire alarm, guard tour, sprinkler waterflow, and sprinkler supervisory systems, see Article 760.

(FPN No. 2): For installation requirements of optical fiber cables, see Article 770.

Reprinted with permission from NFPA 70-1999.

For installation requirements for network-powered broadband communications circuits, see Article 830.

Reprinted with permission from NFPA 70-1999.

Article 810 for:

> radio and television receiving equipment and amateur radio transmitting and receiving equipment, but not equipment and antennas used for coupling carrier current to power line conductors.
>
> Reprinted with permission from NFPA 70-1999.

and Article 820 for:

> coaxial cable distribution of radio frequency signals typically employed in community antenna television (CATV) systems.
>
> (FPN): Where the installation is other than coaxial, see Articles 770 and 800 as applicable.
>
> Reprinted with permission from NFPA 70-1999.

A new Article 830 has been added for "Network Powered Broadband Communications Circuits." This is a new generation communication system and will provide multimedia services in any combination of audio, voice, video, data, and interactive services. This new system may operate at higher levels of power and recognizes two levels of power.

Network Power Source	Low	Medium
Maximum power rating (volt-amperes)	100	100
Maximum voltage rating (volts)	100	150

Chapter 8 does not reference Table 300-5 (wiring method burial requirements). Therefore, the table cannot be applied. There are no burial depth requirements for these communication cables. The reason for that is that the *NEC®* is primarily a safety document. Communications generally pose no safety hazards by interruption of service. Therefore, the installers often bury these cables for service and economic reasons. However, they pose no safety hazards, so the *NEC®* does not set any requirements. Local requirements or amendments may apply.

Sections 800-6 and 820-6 cover the mechanical execution of work for communications circuits, and equipment shall be installed in a neat and workmanlike manner. Cables shall be supported by the building structure in such a manner that the cable will not be damaged by normal building use.

Chapter 8 is unique. Study it carefully. Study the scope carefully, as shown above each article, to be sure you are answering a question from the right article and section of this chapter.

CHAPTER 13 QUESTION REVIEW

> Each lesson is designed purposely to require the student to apply the entire *NEC®* text, not specific chapters or articles. It has been found that when studying for a timed, open-book examination the student must gain proficiency in the Table of Contents, the Index, and the ability to move quickly from cover to cover to find the correct answers to each question in a timely fashion.

1. Does the Code address grounding CATV service to a dwelling unit?

   Answer: _____

   _____

   _____

   _____

   Reference: _____

2. After making an installation of a satellite dish in a local residence, the inspector informs you that you must bury the cable from the satellite dish to the house in accordance with Table 300-5 of the *NEC®*. Was that inspector correct?

   Answer: _____

   _____

   _____

   _____

   Reference: _____

3. Does the Code require grounding the metal sheath of CATV cable entering a dwelling?

   Answer: _____

   _____

   _____

   _____

   Reference: _____

4. According to the *National Electrical Code®*, open conductors for communication equipment on a building shall be separated at least ____ feet from *lightning* conductors, where practicable.

   Answer: _____

   _____

   _____

   _____

   Reference: _____

5. In a highly hazardous location where propane is used, the customer states that the wiring to the control circuit is of sufficient low voltage that a spark cannot be generated from this circuit. The devices and wiring have a control drawing number on them. What type wiring is most likely employed in this installation where there is not enough voltage to generate a spark sufficient to cause an explosion?

Answer: _____

_____

_____

_____

Reference: _____

6. You recently encountered a receptacle with an orange triangle on the face. You were told that this was an isolated ground receptacle. Where can this be found in the *National Electrical Code®*?

Answer: _____

_____

_____

_____

Reference: _____

7. In wiring a large church, an installation of a pipe organ was to be required. Are pipe organs covered in the *NEC®*?

Answer: _____

_____

_____

_____

Reference: _____

8. In making an installation, it was noted the clearance between two large switchboards was rather tight, only about 3 feet, although with the doors open there was not room to pass. Are there any requirements in the *National Electrical Code®* to cover the requirements for working clearances?

Answer: _____

_____

_____

_____

Reference: _____

9. After completing the wiring of a new home, the inspector came and told you to cover the panelboard, that if the sheetrockers got their spackling material or paint on the panel, it was a code violation. Is this correct?

   Answer: _____

   _____

   _____

   _____

   Reference: _____

10. You recently mounted a new service on a new mobile home on a privately owned lot. The inspector turned it down and said this was not permitted. Which section of the *National Electrical Code®* did you violate?

    Answer: _____

    _____

    _____

    _____

    Reference: _____

11. What is an optional standby system?

    Answer: _____

    _____

    _____

    _____

    Reference: _____

12. Is color-coding required in the *NEC®*? You know that the neutral conductor is required to be white and the grounding conductor is required to be green, but what about the other ungrounded conductors?

    Answer: _____

    _____

    _____

    _____

    Reference: _____

13. In making an installation to an outdoor ballfield, you proposed a common neutral to be run with multiple branch-circuit conductors. How many branch-circuit conductors can be supplied with a common neutral?

    Answer: _____

    _____

    _____

    _____

    Reference: _____

14. You have been asked to wire a dairy barn. Are there any specific requirements in the *NEC®* for this type installation?

Answer: _____

_____

_____

_____

Reference: _____

15. You recently observed a local electrician installing wiring under a carpet in a bank. Is this wiring method permitted? If so, which article covers wiring under carpets?

Answer: _____

_____

_____

_____

Reference: _____

16. More buildings are increasingly requiring lightning arrestors to be installed on the service. Which section of the *National Electrical Code®* covers this installation?

Answer: _____

_____

_____

_____

Reference: _____

17. Are temporary wiring methods covered in the *National Electrical Code®*? You regularly install a temporary meter loop to supply power to the construction crews until the permanent power system has been installed. Where is this covered in the *NEC®*?

Answer: _____

_____

_____

_____

Reference: _____

18. A 480-volt, 3-phase circuit with a rated load of 150 kW draws 210 amperes. The power factor is _____ %.

Answer: _____

_____

_____

_____

Reference: _____

19. A _____-horsepower, 208-volt, 3-phase motor operating at 91% efficiency has an input current of 141 amperes (Hp = 746 watts).

    Answer: _____

    _____

    _____

    _____

    Reference: _____

20. A feeder supplying a second building requires a maximum conductor resistance of 0.02 ohms. The minimum size conductor must have a resistance of no more than _____ ohms per 1,000 feet. The total length of the circuit conductor is 190 feet.

    Answer: _____

    _____

    _____

    _____

    Reference: _____

Chapter Fourteen

ELECTRICAL AND *NEC*® QUESTION REVIEW

Now you are about to begin the final chapter of this book, Chapter 14. You should have a good understanding of the Code arrangement and its content. You should now understand that the chapters are arranged in the order in which you use them in most common installations, from Article 90, Introduction; Chapter 1, general requirements; Chapter 2, wiring and protection; Chapter 3, wiring methods and materials; Chapter 4, general use equipment; Chapters 5, 6, and 7, special occupancies, uses and equipment; and Chapter 8, communication systems. The final chapter in the *NEC*® is Chapter 9, which has the tables and examples. Furthermore, you should now understand this arrangement and know the articles contained in each chapter. For instance, you should understand that the article covering rigid metal conduit would automatically be found in Chapter 3, because rigid metal conduit is a wiring method. You should understand that the conditions for installation of a light fixture would be found in Chapter 4, because light fixtures are general use equipment. Or you should understand that hospital installation requirements would be found in Chapter 5, because a hospital is a special occupancy and those are found in Chapter 5, such as mobile homes, hazardous materials and spray booths, commercial garages, and so on. You should automatically remember that a sign would be found in Chapter 6, special equipment, because although signs are common, they are not general equipment found in all types of installations. Signs are special equipment and are found in Chapter 6. These things must be understood before you take the examination. It is necessary so that you might use the *National Electrical Code*® efficiently as a reference document. Memorization is not recommended. However, with a good, clear understanding, one can refer to the Table of Contents in the front of the book and quickly find the article number if you have an understanding of where that article should be found in the Code. You should then be able to go to the Index and find the proper reference so you might quickly look up the question and prove the answer.

Once this has been done, you should have no problem passing any competency exam given throughout the country that is based on your knowledge of the *National Electrical Code*®.

With that competency achieved and your understanding of the *National Electrical Code*® arrangement ensured, the next important task you must improve is the understanding of what each question is asking. You must read the question through, then determine exactly what the question is asking. You will find key words, and with those key words you can simplify the question generally so it can be quickly researched using the Table of Contents, the Index, and your knowledge of the Code arrangement.

Example: As an example, in the following question, select the key words and simplify the question:

"May Type NM cable be used for a 120/240-volt branch-circuit as temporary wiring in a building under construction where the cable is supported on insulators at intervals of not more than 10 feet?"

Now, you may rephrase the question slightly different in your mind, or you may look for the key words. NM cable is a key word, temporary wiring is a key, and supported on insulators at intervals of not more than 10 feet is a key. So,

"Is NM cable permitted as temporary wiring where supported on insulators?"

Now, with the three key words in mind, we must remember that they are not asking where this wiring method is acceptable, they are telling you that the installation would be temporary wiring. If we go to the Table of Contents in the front of the *NEC*®, we find that temporary wiring is found in Article 305. As we turn to Article 305 and we look in the scope under 305-1, we find out that, yes, lesser methods are acceptable. As we research that article quickly, we find in Section 305-4(c) that this open wiring cable is an acceptable method, provided that it is supported on insulators at intervals of not more than 10 feet, so the answer would be yes. References would be Article 305, Sections 305-1 and 305-4(c). As you see in this case when selecting the key words or phrases, using the Table of Contents was sufficient and it was not necessary to go to the Index.

Example: Let us look at another example. The question is:

"Is liquidtight flexible nonmetallic conduit permitted to be used in circuits in excess of 600 volts?"

As we look at that question, we see that liquidtight flexible nonmetallic conduit is the key phrase. Then we look at permitted use as a key phrase and 600 volts as the final key phrase.

Again, we can go to the Table of Contents and find that liquidtight flexible nonmetallic conduit is found in Article 351, Part B. As we turn to Article 351, Part B, we see uses permitted in Section 351-23(a), uses not permitted in Section 351-23(b).

Immediately as we scan down the uses permitted and the uses not permitted, we find that in Section 351-23(b)(4) that liquidtight flexible nonmetallic conduit is not permitted for circuits in excess of 600 volts. However, you will note there is an exception for electrical signs, so the answer is: "Yes, for electrical signs by exception. No, generally."

If you will take each question and break it down so you are looking at the key phrase or key word, you will find it much easier to reference the correct Code article and section, and you will not waste the valuable time needed to complete your examination. All questions will not be as easy as these two examples to identify the key phrases, but in all questions there are key phrases or key words and with the practice you will receive during the following exercises in Chapter 14, you should greatly improve your competency and speed when preparing for an examination, which will allow you ample time to get through the examination and still have some time to do additional research on the tough questions. What you will find by following these examples is that there will be fewer tough questions in the exam. Good luck!

PRACTICE EXAM 1

> **Each lesson is designed purposely to require the student to apply the entire *NEC*® text, not specific chapters or articles. It has been found that when studying for a timed, open-book examination the student must gain proficiency in the Table of Contents, the Index, and the ability to move quickly from cover to cover to find the correct answers to each question in a timely fashion.**

1. Are all 125-volt, 15- and 20-ampere receptacles in the service area of a commercial garage required to be protected by a GFCI?

   Answer: _____

   _____

   _____

   _____

   Reference: _____

2. A lighting fixture is installed over a hydromassage bathtub. Is GFCI protection required for the fixture?

   Answer: _____

   _____

   _____

   _____

   Reference: _____

3. A nongrounding-type receptacle is to be replaced with a GFCI receptacle because a grounding means does not exist within the box. Can you supply downstream receptacles from the GFCI? Are these downstream receptacles required to be 2-wire, nongrounding type?

   Answer: _____

   _____

   _____

   _____

   Reference: _____

4. A kitchen range has a built-in 125-volt receptacle and is within 6 feet of the sink. Does the Code require such a receptacle to be protected with a GFCI?

   Answer: _____

   _____

   _____

   _____

   Reference: _____

5. Does the *NEC®* prohibit the use of nonmetallic outlet boxes with metal raceways?

   Answer: _____

   _____

   _____

   _____

   Reference: _____

6. Does the *NEC®* permit lighting fixtures to be installed on trees?

   Answer: _____

   _____

   _____

   _____

   Reference: _____

7. Electrical nonmetallic tubing (ENT) can be used concealed within walls, ceilings, and floors where the walls, ceilings, and floors provide a thermal barrier of material that has at least _____.

   Answer: _____

   _____

   _____

   _____

   Reference: _____

8. Does the Code require any extra protection for electrical nonmetallic tubing (ENT) when it is installed through openings in metal studs?

   Answer: _____

   _____

   _____

   _____

   Reference: _____

9. What are the strapping requirements for ENT when installed in metal studs?

   Answer: _____

   _____

   _____

   _____

   Reference: _____

10. Does the Code require any extra protection for Type NM cable when it is installed through openings in metal studs?

    Answer: _____

    _____

    _____

    _____

    Reference:_____

11. When UF cable is used for interior wiring, are the conductors required to be rated at 90°C?

    Answer: _____

    _____

    _____

    _____

    Reference:_____

12. Must "hospital grade" receptacles be installed throughout hospitals?

    Answer: _____

    _____

    _____

    _____

    Reference:_____

13. When railings or lattice work are used for room dividers, does the Code require receptacles to be installed as if they were solid walls?

    Answer: _____

    _____

    _____

    _____

    Reference:_____

14. In an apartment project, multiwire branch-circuits are routed through outlet boxes to supply receptacles. Can the neutral conductor "continuity be assured" by two connections to the screw terminals of the receptacle?

    Answer: _____

    _____

    _____

    _____

    Reference:_____

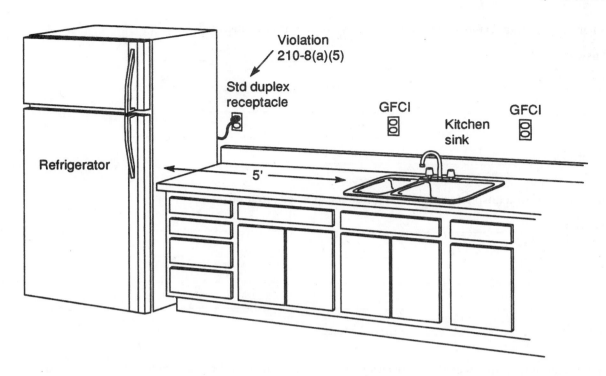

15. A small appliance circuit receptacle is above the counter next to the refrigerator. The refrigerator is plugged into this receptacle that is 5 feet from the kitchen sink. Is the receptacle required to have GFCI protection?

Answer: _____

_____

_____

_____

Reference: _____

16. Does the Code require outdoor receptacles to be installed on balconies of high-rise apartment buildings?

Answer: _____

_____

_____

_____

Reference: _____

17. The Code did require that switches and circuit breakers used as switches be installed so the center of the grip of the operating handle when in its highest position was no more than 6 feet above the floor or working platform. Is this still the case?

Answer: _____

_____

_____

_____

Reference: _____

18. When can lighting fixtures be used as raceways for circuit conductors?

    Answer: _____

    _____

    _____

    _____

    Reference: _____

19. In what areas of a hospital are hospital grade receptacles required?

    Answer: _____

    _____

    _____

    _____

    Reference: _____

20. Is it a requirement that receptacles over a wet bar be spaced according to cabinet or floor line requirements?

    Answer: _____

    _____

    _____

    _____

    Reference: _____

21. A weatherproof receptacle cover must be weatherproof when in use (attachment cap inserted) or shall have a _____.

    Answer: _____

    _____

    _____

    _____

    Reference: _____

22. Would electrical nonmetallic tubing be an acceptable wiring method for a wet-niche lighting fixture or some portion of the circuit?

    Answer: _____

    _____

    _____

    _____

    Reference: _____

23. Can nonmetallic sheathed cable be used to supply a recessed fluorescent fixture?

    Answer: _____

    _____

    _____

    _____

    Reference: _____

24. Do conductors for festoon lighting have to be rubber covered?

    Answer: _____

    _____

    _____

    _____

    Reference: _____

25. Can an electric discharge fixture such as a fluorescent strip be used as a raceway for circuit conductors?

    Answer: _____

    _____

    _____

    _____

    Reference: _____

PRACTICE EXAM 2

1. Are all fixtures located in agriculture buildings to be listed as dusttight or watertight?

   Answer: _____

   _____

   _____

   _____

   Reference: _____

2. When installing AC cable horizontally through metal studs, are insulators needed?

   Answer: _____

   _____

   _____

   _____

   Reference: _____

3. What types of cable assemblies are acceptable for use as feeders to a floating building?

   Answer: _____

   _____

   _____

   _____

   Reference: _____

4. Would nonmetallic sheathed cable be acceptable to wire a pool-associated motor?

   Answer: _____

   _____

   _____

   _____

   Reference: _____

5. Would liquidtight flexible conduit be acceptable for enclosing conductors from an adjacent weatherproof box to an approved swimming pool junction box?

   Answer: _____

   _____

   _____

   _____

   Reference: _____

6. Is the cord- and plug-connection of an electric water heater acceptable under the *NEC*®? If so, would a standard dryer cord be acceptable (240 volts with neutral)? (Assume ampacities, and so on, were acceptable under the *NEC*®.)

   Answer: _____

   _____

   _____

   _____

   Reference: _____

7. Does the Code permit motor controllers and disconnects to be used as junction boxes or wireways?

   Answer: _____

   _____

   _____

   _____

   Reference: _____

8. In determining the locked-rotor kVA per horsepower of dual voltage electric motor, which voltage must be considered?

   Answer: _____

   _____

   _____

   _____

   Reference: _____

9. Would electrical nonmetallic tubing be acceptable to enclose the insulated conductors for a panelboard (not part of service equipment) feeding pool-associated equipment?

   Answer: _____

   _____

   _____

   _____

   Reference: _____

10. Does the Code permit the installation of 15-ampere-rated receptacles on the 20-ampere small appliance branch-circuits required in a dwelling?

    Answer: _____

    _____

    _____

    _____

    Reference: _____

11. Does the Code permit the installation of a 15-ampere-rated duplex receptacle on a 20-ampere separate circuit for a microwave oven with a nameplate reading of 13 amperes?

    Answer: _____

    _____

    _____

    _____

    Reference:_____

12. What does the term conduit in free air mean at the top of Table 310-17?

    Answer: _____

    _____

    _____

    _____

    Reference: _____

13. Has the *National Electrical Code*® restricted nonmetallic conduit (PVC) from being installed in cold weather areas?

    Answer: _____

    _____

    _____

    _____

    Reference:_____

14. All temporary wiring 15A and 20A, 125-volt receptacles not a part of the permanent building structure need to have GFCI protection. Does this mean that if a permanently installed receptacle is installed and energized that no GFCI protection is required even though this outlet can be used by construction personnel?

    Answer: _____

    _____

    _____

    _____

    Reference:_____

15. In the Code, would a GFCI-protected receptacle be required on a wet bar in a dwelling?

    Answer: _____

    _____

    _____

    _____

    Reference:_____

16. Must the garbage disposal receptacle located under the sink be GFCI protected because it is within 6 feet of the sink?

    Answer: _____

    _____

    _____

    _____

    Reference: _____

17. What is the maximum number of duplex receptacles permitted on a 2-wire, 15 amp circuit in a dwelling?

    Answer: _____

    _____

    _____

    _____

    Reference: _____

18. How many receptacles are required at the patient's bed location in a critical care facility?

    Answer: _____

    _____

    _____

    _____

    Reference: _____

19. A ¼ horsepower circulating pump is to be located in a closet space housing a gas-fired water heater in a large single-family residence. Can the circulating pump be cord and plug connected?

    Answer: _____

    _____

    _____

    _____

    Reference: _____

20. In lieu of switching a ceiling light, the Code permits switching a receptacle. Can the switched receptacle be counted as one of the receptacles required for the 12-foot rule?

    Answer: _____

    _____

    _____

    _____

    Reference: _____

21. A laundry room in a single-family residence has a receptacle outlet on one wall and a second receptacle on the opposite wall. Can these two outlets be fed by the same laundry branch outlet?

Answer: _____

_____

_____

_____

Reference: _____

22. Antique stores and others are selling hanging fixtures without a grounding conductor. Are these fixtures in violation of the *NEC*®?

Answer: _____

_____

_____

_____

Reference: _____

23. Can 6 feet of flexible metal conduit be installed between "rigid nonmetallic" raceways used for service-entrance conductors?

Answer: _____

_____

_____

_____

Reference: _____

24. Many kitchen and bathroom sinks, tubs, and showers are provided with short sections of metal pipe (6 inches to 2 feet) for connection to nonmetallic pipe systems. Is it required to bond the several short sections of metal pipe to the service equipment?

Answer: _____

_____

_____

_____

Reference: _____

25. Is there a limit on the number of extension boxes that can be joined together?

Answer: _____

_____

_____

_____

Reference: _____

PRACTICE EXAM 3

1. A switch box was installed in a floor area. A receptacle outlet with a standard plate was installed on the box. Does this standard outlet box and receptacle installed in the floor comply with the Code?

    Answer: _____

    _____

    _____

    _____

    Reference: _____

2. A 4 inch × 2⅛ inch square NM box is marked with the number of No. 14, 12, and 10 conductors that can be installed in it. Does this mean that No. 8 conductors are not permitted in the box?

    Answer: _____

    _____

    _____

    _____

    Reference: _____

3. Are outdoor receptacle outlets required at each dwelling unit of a multifamily dwelling?

    Answer: _____

    _____

    _____

    _____

    Reference: _____

4. Do portable lamps wired with flexible cord require a polarized attachment plug?

    Answer: _____

    _____

    _____

    _____

    Reference: _____

5. When figuring receptacles for a bedroom in a residence, must you include the space behind the door?

    Answer: _____

    _____

    _____

    _____

    Reference: _____

6. Are metal boxes required for splices in temporary wiring?

    Answer: _____

    _____

    _____

    _____

    Reference: _____

7. What is the maximum number of No. 12 conductors permitted in a 4 inch × 1½ inch deep octagon box?

    Answer: _____

    _____

    _____

    _____

    Reference: _____

8. When a junction box contains a combination of conductors, such as size No. 14 and No. 12 wires, what table is used to determine the size of the box?

    Answer: _____

    _____

    _____

    _____

    Reference: _____

9. Can you use threaded intermediate metal conduit in a Class 1, Division 1 location?

    Answer: _____

    _____

    _____

    _____

    Reference: _____

10. Does the Code consider EMT as conduit?

    Answer: _____

    _____

    _____

    _____

    Reference: _____

11. Can suspended fluorescent fixtures be connected together with unsupported EMT between them?

    Answer: _____

    _____

    _____

    _____

    Reference: _____

12. Must lay-in fluorescent lighting fixtures be fastened to the grid tee-bars of a suspended ceiling system?

    Answer: _____

    _____

    _____

    _____

    Reference: _____

13. Can fixtures marked "suitable for damp locations" be used in wet locations?

    Answer: _____

    _____

    _____

    _____

    Reference: _____

14. A UL-listed fluorescent fixture is equipped with a Class "P" ballast. Does the Code permit such a fixture to be surface mounted on a combustible low-density cellulose fiberboard ceiling?

    Answer: _____

    _____

    _____

    _____

    Reference: _____

15. Are all recessed incandescent fixtures required to have thermal protection?

    Answer: _____

    _____

    _____

    _____

    Reference: _____

16. Is it necessary to identify the high leg of a 240/120-volt system at the motor disconnect of a 3-phase motor?

    Answer: _____

    _____

    _____

    _____

    Reference: _____

17. Is a GFCI protection required for a receptacle located on a food preparation island that is less than 6 feet from the countertop sink?

    Answer: _____

    _____

    _____

    _____

    Reference: _____

18. Can conductors that pass through a panelboard also be spliced in the panelboard?

    Answer: _____

    _____

    _____

    _____

    Reference: _____

19. Are the enclosures for mercury vapor lighting fixtures installed 8 feet above grade on poles outdoors required to be grounded?

    Answer: _____

    _____

    _____

    _____

    Reference: _____

20. Would an exit light with an emergency pack capable of supplying more than $1\frac{1}{2}$ hours of power have to be connected to an emergency circuit ahead of the main disconnect?

    Answer: _____

    _____

    _____

    _____

    Reference: _____

21. A grounded 20 amp receptacle is installed in a residential garage to serve an air compressor. Would this receptacle require GFCI protection?

    Answer: _____

    _____

    _____

    _____

    Reference: _____

22. If additional receptacles are installed in a bathroom, not adjacent to the wash basin, are they required to be GFCI protected?

    Answer: _____

    _____

    _____

    _____

    Reference: _____

23. Must a receptacle in the kitchen installed for a lighting fixture be GFCI protected?

    Answer: _____

    _____

    _____

    _____

    Reference: _____

24. When installing recessed fixtures (hi-hats) in the ceiling, does Section 370-20 apply to the fixture housing?

    Answer: _____

    _____

    _____

    _____

    Reference: _____

25. What are the support requirements for service-entrance cables?

    Answer: _____

    _____

    _____

    _____

    Reference: _____

PRACTICE EXAM 4

1. An underground cable is routed from the service equipment in the building underground and up a pole in a commercial parking lot to supply an overhead light. What is the minimum burial depth for the cable below the base of the pole?

   Answer: _____

   _____

   _____

   _____

   Reference: _____

2. Does the Code permit 15-ampere duplex receptacles to be installed on a 20-ampere branch-circuit?

   Answer: _____

   _____

   _____

   _____

   Reference: _____

3. We are designing a new hotel with 250 guest rooms. It is our intent to install permanent beds, desks, and dressers in each guest room, bolted to the wall. If receptacles are located in accordance with Section 210-52, the receptacles will be located behind the headboard of the bed and behind the dresser. Are the receptacles required behind these pieces of furniture? Would the number of required receptacles be less if the furniture was not bolted to the wall?

   Answer: _____

   _____

   _____

   _____

   Reference: _____

4. Can a paddle fan be supported only by an outlet box if it weighs less than 35 pounds?

   Answer: _____

   _____

   _____

   _____

   Reference: _____

5. When can lighting fixtures be used as raceways for circuit conductors?

   Answer: _____

   _____

   _____

   _____

   Reference: _____

6. Do recessed incandescent lighting fixtures require thermal protection?

   Answer: _____

   _____

   _____

   _____

   Reference: _____

7. Are medium base lampholders in industrial occupancies permitted on 277-volt lighting circuits?

   Answer: _____

   _____

   _____

   _____

   Reference: _____

8. Can a battery-powered (unit equipment) emergency light be supplied directly from a circuit in a panelboard that supplies normal lighting in the area that will be illuminated by the battery light under emergency conditions?

   Answer: _____

   _____

   _____

   _____

   Reference: _____

9. A single transfer switch is permitted to serve one or more branches of the essential electrical system in a small hospital. What constitutes a small hospital?

   Answer: _____

   _____

   _____

   _____

   Reference: _____

10. Would a 3-phase, 4-wire, 1,200-ampere, 480/277-volt service switch with 900-ampere fuses be required to have ground-fault protection?

   Answer: _____

   _____

   _____

   _____

   Reference: _____

11. Are lighting fixtures operating at less than 15 volts between conductors required to be protected by a GFCI over hot tubs or spas?

   Answer: _____

   _____

   _____

   _____

   Reference: _____

12. Could you use electrical metallic tubing (EMT) in the earth below a concrete slab?

   Answer: _____

   _____

   _____

   _____

   Reference: _____

13. An industrial building will be served by several services of different voltage ratings. Is it required that each system be separately grounded?

   Answer: _____

   _____

   _____

   _____

   Reference: _____

14. In residential occupancies, locating protective equipment to comply with the Code becomes a problem. Can a service panel or distribution panelboard be located over a counter or an appliance (washing machine, dryer) that extends away from the wall?

   Answer: _____

   _____

   _____

   _____

   Reference: _____

15. A surface-mounted fluorescent fixture with an integral receptacle is installed above the countertop in the kitchen of a dwelling and is not more than 5½ feet above the floor. Is this fixture required to be supplied by a 20-ampere circuit as are other kitchen receptacles? If within 6 feet of the sink, is the fixture receptacle required to be protected by GFCI?

Answer: _____

_____

_____

_____

Reference: _____

16. What are the requirements for grounding an agricultural building where livestock is housed?

Answer: _____

_____

_____

_____

Reference: _____

17. Can a disconnect switch or breaker located on the meter pole of a farm installation be used as the service equipment for the residence?

Answer: _____

_____

_____

_____

Reference: _____

18. A Type UF cable feeds a 120-volt yard light location on residential property. The circuit is protected by a 15-ampere overcurrent protection device. It is not GFCI protected. What is the minimum burial depth permitted for the cable?

Answer: _____

_____

_____

_____

Reference: _____

19. Can a fixed storage-type water heater having a capacity of 120 gallons or less be cord and plug connected?

Answer: _____

_____

_____

_____

Reference: _____

20. A TV dish antenna is located in the side yard of a single-family dwelling. (a) What is the minimum depth required by the *NEC®* for the coaxial signal cable? (b) How are these antenna units required to be grounded?

Answers: _____

_____

_____

_____

Reference: _____

21. Does the Code permit low-voltage control cables to be supported from the conduit containing the power circuit conductors feeding an air-conditioning unit?

Answer: _____

_____

_____

_____

Reference: _____

22. Is it true that if the underground metallic water piping is not at least 10 feet long, the underground piping system is not adequate by itself as a grounding electrode?

Answer: _____

_____

_____

_____

Reference: _____

23. Does the lighting outlet in the attic have to be switch controlled?

Answer: _____

_____

_____

_____

Reference: _____

24. With regards to recessed lighting fixtures installed over showers, is it permissible to use a gasketed vapor-proof trim only to comply with the Code or must the entire fixture be rated for a wet location?

    Answer: _____

    _____

    _____

    _____

    Reference: _____

25. May overhead outside branch-circuit conductors of No. 10 copper to be used in an unsupported open span up to 50 feet? Is No. 10, listed copper multiconductor UF cable suitable for such an application?

    Answer: _____

    _____

    _____

    _____

    Reference: _____

PRACTICE EXAM 5

1. Is a nonmetallic cable tray permitted by the *NEC®*?

   Answer: _____

   _____

   _____

   _____

   Reference: _____

2. The Code requires clearance for lighting fixtures installed in clothes closets. (a) What are the clearance requirements for surface mounted fixtures mounted on the wall above the door? (b) What are clearance requirements for surface mounted fixtures mounted on the ceiling? (c) What are the clearance requirements for recessed incandescent or fluorescent fixtures ceiling mounted?

   Answers: _____

   _____

   _____

   _____

   Reference: _____

3. Are No. 16 fixture wires counted for the number of conductors in outlet boxes to determine the box fill?

   Answer: _____

   _____

   _____

   _____

   Reference: _____

4. Are switchboards and control panels rated 1,200 amperes or more, 600 volts, and over 6 feet wide required to have one entrance at each end of the room?

   Answer: _____

   _____

   _____

   _____

   Reference: _____

5. What size grounding-electrode conductor is required where a grounding conductor is routed to a driven rod and from the rod to a concrete-encased electrode?

   Answer: _____

   _____

   _____

   _____

   Reference: _____

6. Can handle locks on circuit breakers be locked so the power to loads such as emergency lighting, sump pumps, alarm warning circuits, and other types of equipment cannot be cut off by mistake? What sections of the *NEC*® would apply?

   Answer: _____

   _____

   _____

   _____

   Reference:_____

7. Can a service be mounted on a mobile home if the home is designed to be installed on a foundation?

   Answer: _____

   _____

   _____

   _____

   Reference:_____

8. Is an insulated equipment ground required for a pool panelboard feeder fed from the service equipment?

   Answer: _____

   _____

   _____

   _____

   Reference:_____

9. Would the Code allow the use of stainless steel ground rods?

   Answer: _____

   _____

   _____

   _____

   Reference:_____

10. The Code now allows flexible metal conduit for services. What about liquidtight flexible conduit, metallic or nonmetallic?

   Answer: _____

   _____

   _____

   _____

   Reference:_____

11. The grounding conductor can be connected to the grounding electrode by listed connectors, clamps, or other listed means and _____?

Answer: _____

_____

_____

_____

Reference:_____

12. Does the Code require a ground wire in all flexible raceways?

Answer: _____

_____

_____

_____

Reference:_____

13. Is AC cable a permitted wiring method in places of assembly?

Answer: _____

_____

_____

_____

Reference:_____

14. Does the obelisk notation in the footnotes to Tables 310-16 and 17 apply to motor circuits?

Answer: _____

_____

_____

_____

Reference:_____

15. Can liquidtight flexible nonmetallic conduit be used on residential work?

Answer: _____

_____

_____

_____

Reference:_____

16. Are there any Code requirements against installing a bare neutral wire inside the service conduit?

Answer: _____

_____

_____

_____

Reference: _____

17. Can the branch-circuit and Class II control wires for two 3-phase motors be installed in a common raceway of the proper size?

Answer: _____

_____

_____

_____

Reference: _____

18. The Code explicitly requires bonding around concentric knockouts on service equipment, but does the Code require that equipment downstream from the service equipment be bonded in the same manner?

Answer: _____

_____

_____

_____

Reference: _____

19. Is nonmetallic sheathed cable (Romex) permitted to run through cold air returns if it is sleeved with thin-wall conduit or greenfield in short lengths?

Answer: _____

_____

_____

_____

Reference: _____

20. Is it necessary to pigtail the neutral when connecting receptacles to a multiwire circuit?

Answer: _____

_____

_____

_____

Reference: _____

21. When a motor-control circuit extends beyond the control enclosure, what is the maximum overcurrent protection acceptable for No. 12 AWG copper control wires?

    Answer: _____

    _____

    _____

    _____

    Reference: _____

22. Is an equipment bonding conductor required to bond a cover-mounted receptacle to a surface-mounted outlet box?

    Answer: _____

    _____

    _____

    _____

    Reference: _____

23. When computing branch-circuit and feeder loads, what is the normal system voltages that shall be used?

    Answer: _____

    _____

    _____

    _____

    Reference: _____

24. Is it permissible to run four No. 12 AWG THW copper conductors in a cable or raceway where they encounter 100°F ambient temperatures to supply cord- and plug-connected loads protected at 20 amperes?

    Answer: _____

    _____

    _____

    _____

    Reference: _____

25. When flat conductor cable, Type FCC, is installed under carpet, what is the maximum size carpet squares permitted?

    Answer: _____

    _____

    _____

    _____

    Reference: _____

PRACTICE EXAM 6

1. Are bonding jumpers required on metal feeder and branch-circuit raceways containing circuits of more than 250 volts to ground where oversize concentric or eccentric knockouts are encountered?

    Answer: _____

    _____

    _____

    _____

    Reference: _____

2. What is the maximum number of service disconnects allowed for a set of service-entrance conductors installed in an apartment for the Code?

    Answer: _____

    _____

    _____

    _____

    Reference: _____

3. (a) Is a GFCI required on or within an outdoor portable sign? (b) Would a GFCI-protected branch-circuit supplying the sign be acceptable?

    Answers: _____

    _____

    _____

    _____

    Reference: _____

4. Is a lighting outlet required in the crawl space of a manufactured building if equipment requiring service is installed in the space?

    Answer: _____

    _____

    _____

    _____

    Reference: _____

5. Can an office partition assembly be cord and plug connected?

    Answer: _____

    _____

    _____

    _____

    Reference: _____

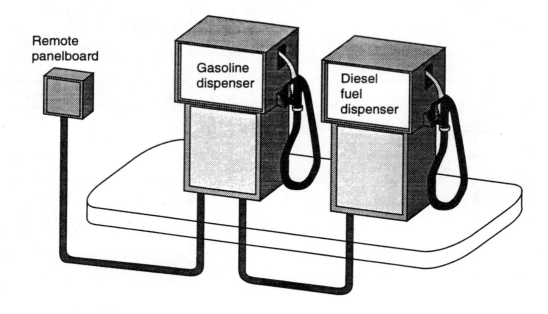

6. In the figure above, a diesel fuel dispenser is mounted adjacent to a gasoline dispenser on a service station island. A sealing fitting is installed in the conduit entering the gasoline dispenser. Is a sealing fitting required in the conduit entering the diesel fuel dispenser?

Answer: _____

_____

_____

_____

Reference:_____

7. In residential-type occupancies, does the wall space behind doors where the door opens against the wall count when spacing the receptacles in a room?

Answer: _____

_____

_____

_____

Reference:_____

8. A four-family apartment building is supplied by an overhead drop to a 300 kcmil THW aluminum riser. Four electric meters are grouped and have 100-ampere breakers below the meters to supply a panel in each apartment. Number 3 THW copper feeders are run to each apartment unit. An individual water meter is provided for each apartment. Each water meter is supplied with nonmetallic pipe. The water system for each apartment is copper above grade. (a) What size grounding electrode conductor is required? (b) How would this conductor be attached to the metallic water piping in each apartment? (c) Is a supplementary driven electrode required?

Answers:_____

_____

_____

_____

Reference:_____

9. Is a No. 14 copper conductor allowed in a kitchen of a dwelling unit if it serves a single fixed appliance such as a dishwasher?

Answer: _____

_____

_____

_____

Reference:_____

10. Can UF-type cable be used to supply a swimming pool motor?

Answer: _____

_____

_____

_____

Reference:_____

11. Are No. 16 and No. 18 fixture wires counted for the number of conductors in outlet boxes for box fill?

Answer: _____

_____

_____

_____

Reference:_____

12. Is a receptacle outlet required for a wet bar and/or if an outlet is installed within 6 feet of the wet bar sink, would it be required to be GFCI protected?

Answer: _____

_____

_____

_____

Reference: _____

13. When calculating the demand for a 3 kW and a 6 kW range in a dwelling, would you use Table 220-19, Column B for the 3 kW and Column C for the 6 kW range or would you combine them and use either B or C as two ranges?

Answer: _____

_____

_____

_____

Reference: _____

14. When sizing the branch-circuit conductors to a 4.5 kW dryer, would you use 4.5 kW or 5 kW in making the calculations?

Answer: _____

_____

_____

_____

Reference: _____

15. What is the maximum weight of a ceiling fan that can be supported by an outlet box?

Answer: _____

_____

_____

_____

Reference: _____

16. Where does the *NEC®* list or calculate the number of conductors allowed in flexible metal conduit?

Answer: _____

_____

_____

_____

Reference: _____

17. Are ceiling grid support wires that rigidly support a lay-in ceiling acceptable for the sole support of junction boxes above the lay-in ceiling?

    Answer: _____

    _____

    _____

    _____

    Reference: _____

18. Can electrical equipment approved for a Class I location be used in a Class II location?

    Answer: _____

    _____

    _____

    _____

    Reference: _____

19. Can THHN insulated conductors be used outside for service-entrance conductors where exposed to the elements?

    Answer: _____

    _____

    _____

    _____

    Reference: _____

20. Who has the authority to determine acceptability of electrical equipment and materials?

    Answer: _____

    _____

    _____

    _____

    Reference: _____

21. Can more than one receptacle in a laundry room be supplied by the laundry branch-circuit?

    Answer: _____

    _____

    _____

    _____

    Reference: _____

22. Many air conditioner units are being installed on roof tops of apartment buildings for different reasons. Is a receptacle adjacent to such equipment required for servicing this equipment?

Answer: _____

_____

_____

_____

Reference:_____

23. Section 430-72 applies to motor control circuits that are tapped from the load side of motor branch-circuit short-circuit and ground-fault protective devices. Since Column B of Table 430-72(b) covers control circuits that do not extend beyond the motor control equipment enclosure and Column C covers those that do extend beyond, what does Column A cover?

Answer: _____

_____

_____

_____

Reference:_____

24. A 240-volt, single-phase circuit with a rated load of 1.8 kW draws 9 amperes of current. The power factor is _____ %.

Answer: _____

_____

_____

_____

Reference:_____

25. Can you run multioutlet assembly from one room to another room through a sheetrock wall?

Answer: _____

_____

_____

_____

Reference:_____

PRACTICE EXAM 7

1. Illuminated exit signs are required by local building codes, and battery packs will be used. Do these signs have to be installed on a lighting circuit in the area of the sign or could they be on a separate branch-circuit?

   Answer: _____

   _____

   _____

   _____

   Reference: _____

2. Can a metal cable tray be used as an equipment grounding conductor?

   Answer: _____

   _____

   _____

   _____

   Reference: _____

3. Can surface nonmetallic raceway be used outside in a wet location?

   Answer: _____

   _____

   _____

   _____

   Reference: _____

4. Switchboards and control panels rated 1,200 amperes or more, 600 volts or less, and over 6 feet wide are required to have one entrance at each end of the room. What conditions would permit only one entrance to this room?

   Answer: _____

   _____

   _____

   _____

   Reference: _____

5. Is Type MC cable required to have a bushing installed like the bushings used on armored cable?

   Answer: _____

   _____

   _____

   _____

   Reference: _____

6. Can electrical nonmetallic tubing be surface mounted in a warehouse building?

    Answer: _____

    _____

    _____

    _____

    Reference: _____

7. How deep must a residential branch-circuit be buried if it passes under the driveway?

    Answer: _____

    _____

    _____

    _____

    Reference: _____

8. Where more than one building or structure is on the same property and under single management, the Code requires each building to have a disconnecting means at each such building or structure of 600 volts or less. Under what conditions can the disconnecting means be located elsewhere on the premises?

    Answer: _____

    _____

    _____

    _____

    Reference: _____

9. Are cable trays permitted in a grain elevator?

    Answer: _____

    _____

    _____

    _____

    Reference: _____

10. If a fixture is approved for use as a raceway, can this raceway be used for conductors to supply a circuit or circuits beyond the fixture?

    Answer: _____

    _____

    _____

    _____

    Reference: _____

11. What Code section prohibits bare neutral feeders?

Answer: _____

_____

_____

_____

Reference: _____

12. Would a large mercantile store be considered a place of assembly if it can hold more than 100 people?

Answer: _____

_____

_____

_____

Reference: _____

13. Can the neutral be reduced by 70 percent for data processing equipment over 200 amperes?

Answer: _____

_____

_____

_____

Reference: _____

14. What ampacity correction factor would you use for No. 12 THW copper conductors used in an ambient temperature of 78°F? (Not more than three conductors in raceway, in free air, 240-volt circuit.)

Answer: _____

_____

_____

_____

Reference: _____

15. How are box volumes calculated when the box contains different size conductors and a combination of clamps, devices, and studs?

Answer: _____

_____

_____

_____

Reference: _____

16. What is the minimum service-drop clearance required over a residential yard where the drop conductors do not exceed 150 volts to ground?

    Answer: _____

    _____

    _____

    _____

    Reference: _____

17. An equipment room containing electrical, telephone, and air-handling equipment uses the space in the room for air-handling purposes. Is this room considered a plenum?

    Answer: _____

    _____

    _____

    _____

    Reference: _____

18. As applied to lighting in a clothes closet, is the entire 12-inch or width of shelf area required to be unobstructed to the floor or could part of this area be used for storage?

    Answer: _____

    _____

    _____

    _____

    Reference: _____

19. Would an outdoor lighting fixture be considered as grounded by the physical connection to a grounded metal support pole or would an equipment grounding conductor connection directly to the fixture be required?

    Answer: _____

    _____

    _____

    _____

    Reference: _____

20. Would entrance-type lighting fixtures that are installed under roof overhangs, porch roofs, or canopies be required to be suitable for damp locations?

    Answer: _____

    _____

    _____

    _____

    Reference: _____

21. Table 430-147 through 430-150 list the horsepower and the full-load currents of motors. How could you determine the full-load current of a motor not listed, such as a 35-horsepower, 460-volt motor?

Answer: _____

_____

_____

_____

Reference: _____

22. Where a single NM (Romex) cable is pulled into a raceway, what percent may the cross section of the conduit be filled?

Answer: _____

_____

_____

_____

Reference: _____

23. Can you use Note 3 to 0–2000 Volt Ampacity Tables for a feeder to each apartment in a building regardless of whether the mains are in the apartment units or on the outside of the building?

Answer: _____

_____

_____

_____

Reference: _____

24. Can swimming pool equipment such as a pump motor be calculated as a fixed appliance when applying 220-17 to a dwelling?

Answer: _____

_____

_____

_____

Reference: _____

25. Is a GFCI receptacle required in a detached garage of a dwelling?

Answer: _____

_____

_____

_____

Reference: _____

PRACTICE EXAM 8

1. Are all recessed lighting fixtures that are to be installed in a suspended ceiling required to be thermally protected?

   Answer: _____

   _____

   _____

   _____

   Reference: _____

2. How can you ground an agricultural building and comply with the Code?

   Answer: _____

   _____

   _____

   _____

   Reference: _____

3. Is Table 220-19 applicable to microwave ovens and convection cooking ovens?

   Answer: _____

   _____

   _____

   _____

   Reference: _____

4. Is an autotransformer recognized for use in a motor control circuit?

   Answer: _____

   _____

   _____

   _____

   Reference: _____

5. Can service equipment be mounted on a floating dwelling unit that is moved frequently?

   Answer: _____

   _____

   _____

   _____

   Reference: _____

6. Is it permissible to use neon lighting on or in a dwelling unit?

   Answer: _____

   _____

   _____

   _____

   Reference: _____

7. In a large industrial plant, does the Code allow a branch-circuit over 50 amps to feed several outlets such as to supply power to portable equipment that moves from place to place?

   Answer: _____

   _____

   _____

   _____

   Reference: _____

8. Can nonmetallic raceways be used in health care facilities?

   Answer: _____

   _____

   _____

   _____

   Reference: _____

9. Can a conductor be protected by an overcurrent device sized for its ampacity even though the allowable load current is only 50% of the assigned ampacity after derating?

   Answer: _____

   _____

   _____

   _____

   Reference: _____

10. Does the Code permit single-pole circuit breakers in lighting panelboards to be used as a switch for controlling a lighting circuit as an off and on switch?

    Answer: _____

    _____

    _____

    _____

    Reference: _____

11. When determining the required size box, items such as fittings, cable clamps, hickeys, where combinations of different size conductors are used, what size conductor do you deduct for these fittings?

Answer: _____

_____

_____

_____

Reference: _____

12. Where rigid metal conduit is used as service raceways, are double locknuts permitted for the service equipment required continuity?

Answer: _____

_____

_____

_____

Reference: _____

13. A receptacle outlet is to be added to an existing installation, and NM cable is to be "fished in" the partition wall. Does the Code allow a nonmetallic box without cable clamps to be used for such an installation?

Answer: _____

_____

_____

_____

Reference: _____

14. What size general use switch is required on an air-conditioning compressor, sealed (hermetic-type) when the nameplate states FLA of 100 amperes, 3-phase, 230 volts, LRA 580?

Answer: _____

_____

_____

_____

Reference: _____

15. A fluorescent fixture is supplied with $\frac{3}{8}$ inch flexible conduit 5 feet long. Is an equipment grounding required to ground the fixture?

Answer: _____

_____

_____

_____

Reference: _____

16. What is the full-load current of an electric furnace rated 240 volts, single phase, 10 kW when connected to a 208-volt circuit? (Show calculations.)

    Answer: _____

    _____

    _____

    _____

    Reference: _____

17. Can four 400-watt wet-niche lights be connected to one GFCI for a swimming pool?

    Answer: _____

    _____

    _____

    _____

    Reference: _____

18. How does one determine how many conductors conduit bodies (condulets) are approved for?

    Answer: _____

    _____

    _____

    _____

    Reference: _____

19. How can a recessed fixture approved for direct contact with thermal insulation be identified?

    Answer: _____

    _____

    _____

    _____

    Reference: _____

20. Does the 10-foot tap rule apply only to conductors such as wires and not panelboard bus?

    Answer: _____

    _____

    _____

    _____

    Reference: _____

21. How are working clearances measured between enclosed electrical equipment facing each other across an aisle?

Answer: _____

_____

_____

_____

Reference: _____

22. Can either the disposal, dishwasher, or trash compactor be installed on a 20-ampere kitchen small appliance circuit?

Answer: _____

_____

_____

_____

Reference: _____

23. What is the required horizontal distance between service-drops and swimming pools?

Answer: _____

_____

_____

_____

Reference: _____

24. Can a bare No. 4 copper grounding electrode conductor be buried below a concrete floor slab and is there a depth requirement?

Answer: _____

_____

_____

_____

Reference: _____

25. Where the grounding-electrode conductor connects to the water system on the house side of the water meter, is bonding required around valves with sweated connections? Also, what other equipment is required to be bonded around?

Answer: _____

_____

_____

_____

Reference: _____

PRACTICE EXAM 9

1. Can a flood light and/or receptacle be on the GFCI circuit protecting an underwater light?

   Answer: _____

   _____

   _____

   _____

   Reference: _____

2. For a 600-ampere service where 350 MCM copper conductors are parallel in two conduits and only 3-phase, 3-wire is needed, but the transformer is Y-connected and grounded: (a) Must the grounded conductor be brought into the service? (b) Can the grounded conductor be brought in through a single raceway or is it required to be installed in each of the parallel raceways? (c) What is the maximum size of the grounded conductor?

   Answers: _____

   _____

   _____

   Reference: _____

3. What are the minimum dimensions of a junction box illustrated above with only a U-pull of two 3-inch conduits in one side?

   Answer: _____

   _____

   _____

   Reference: _____

4. With two 42 circuit panelboards bolted together, can the 42 circuits from one panelboard feed through the gutter of the second panelboard if the 40% fill in that gutter is not exceeded?

   Answer: _____

   _____

   _____

   _____

   Reference: _____

5. Can suspended fluorescent fixtures be connected together with unsupported EMT between them if the length of the EMT is less than 3 feet?

   Answer: _____

   _____

   _____

   _____

   Reference: _____

6. (a) Are 3-phase generators permitted to be connected to a three-pole transfer switch with a solid neutral or is a four-pole transfer switch that breaks the neutral required? (b) If either is OK, must a grounding electrode be provided at the generator in both cases?

   Answers: _____

   _____

   _____

   _____

   Reference: _____

7. A 600-ampere underground service with four No. 4/0 conductors per phase paralleled is installed in multiple PVC conduits. Must each of the phase conductors, the neutral and equipment grounding conductor be grouped in each raceway if the conduits are PVC?

   Answer: _____

   _____

   _____

   _____

   Reference: _____

8. The raceway for a 240-volt motor circuit includes both the power circuit and control circuit conductors. The power circuit conductors use 600-volt insulation. Must the control circuit conductors also have 600-volt insulation or would 300-volt insulation be acceptable?

Answer: _____

_____

_____

_____

Reference: _____

9. Does the Code either permit or prohibit a sprinkler head over a switchboard?

Answer: _____

_____

_____

_____

Reference: _____

10. A 2,000-ampere, 3-phase, 480-volt ungrounded service is located 300 feet from the water line within a building. Can the grounding-electrode conductor from this service be connected to the building steel in the vicinity of the service, and then 300 feet away bond the steel to the water line such that the building steel is being used as the grounding-electrode conductor?

Answer: _____

_____

_____

_____

Reference: _____

11. Four No. 12 THNN conductors in one conduit supply a continuous lighting load. What is the maximum permissible circuit breaker size and permissible load allowed on each conductor?

Answer: _____

_____

_____

_____

Reference: _____

12. A receptacle outlet is located in a soffit or roof overhang for connection of Christmas decorations or roof heating cable at a dwelling. When this receptacle is located out of reach from the grade, approximately 9 feet above grade, is this receptacle required to be GFCI protected?

Answer: _____

_____

_____

_____

Reference: _____

13. Does the *NEC*® permit more than one NM cable under one device box cable clamp?

Answer: _____

_____

_____

_____

Reference: _____

14. A 400-ampere service uses two 200-ampere panelboards, each with a main breaker. Additional load is added so that one more 100-ampere service disconnect is required. Must the service-entrance conductors be increased in size? (The calculated load is not increased.)

Answer: _____

_____

_____

_____

Reference: _____

15. Can an attachment plug be used as a disconnecting means for a 3-horsepower motor?

Answer: _____

_____

_____

_____

Reference: _____

16. Where in the Code does it permit using No. 4 copper THW conductors for service conductors supplying a 100-ampere service?

Answer: _____

_____

_____

_____

Reference: _____

17. In three-way and four-way switch circuitry in metal raceway, is the neutral required to be in the same raceway with the travelers (switch legs)?

Answer: _____

_____

_____

_____

Reference: _____

18. Single-conductor UF cable comes underground from a submersible pump through a building wall to the pump controller. Are these conductors permitted to be run inside the building?

Answer: _____

_____

_____

_____

Reference: _____

19. In commercial garages, can regular receptacles and plates be used above 18 inches of the floor? What are the wiring restrictions in a garage that has a minimum of four air changes per hour?

Answer: _____

_____

_____

_____

Reference: _____

20. A single-family dwelling has two electric dryer receptacle outlets wired with two No. 10 black conductors to each one and one No. 10 white conductor neutral. Each outlet has a separate 30-ampere circuit breaker. Is one No. 10 neutral conductor permitted to be used with this installation?

Answer: _____

_____

_____

_____

Reference: _____

21. Can ⅜ inch flexible conduit be used on machinery and boilers for limit switches, flow switches, can switches, and solenoids?

Answer: _____

_____

_____

_____

Reference: _____

22. Article 100, Definitions, states that the definition of a building is a structure that stands alone or that is cut off from adjoining structures by fire walls with all openings therein protected by approved fire doors. Could this definition apply to a building with fire-rated floors and all elevator shafts and stairways protected by approved fire rating and fire doors? In other words, could each floor be considered a separate building?

    Answer: _____

    _____

    _____

    _____

    Reference:_____

23. Can unit equipment used for emergency lighting be directly connected on the branch-circuit rather than using a plug-in receptacle?

    Answer: _____

    _____

    _____

    _____

    Reference:_____

24. Section 517-20 specifies GFCI receptacles for wet locations that by definition in Article 100 include outside areas exposed to the weather. Are receptacles required to be GFCI protected when outdoors at health care facilities?

    Answer: _____

    _____

    _____

    _____

    Reference:_____

25. Does a gas furnace in a dwelling require a disconnect switch on the furnace? May the furnace be directly connected onto a basement lighting circuit?

    Answer: _____

    _____

    _____

    _____

    Reference:_____

PRACTICE EXAM 10

1. Are open tube fluorescent fixtures acceptable in a basement garage under an apartment building?

   Answer: _____

   _____

   _____

   _____

   Reference: _____

2. Hot tubs are popular. Are there size and depth requirements to classify them as a pool or are they classified as a pool regardless of size for wiring of circulation pumps, heaters, and so forth?

   Answer: _____

   _____

   _____

   _____

   Reference: _____

3. Rigid metal conduit runs in commercial garage floor and extends up through 18-inch hazardous area with no couplings or fittings in that 18-inch area. Is a seal-off fitting required in this conduit?

   Answer: _____

   _____

   _____

   _____

   Reference: _____

4. Three single-phase, 480-volt transformers rated at 104 amperes each are connected in a wye bank for a 3-phase system. What is the maximum overcurrent protection permitted for the primary when the secondary is also protected? Would the primary protection be different if connected in a delta bank?

   Answer: _____

   _____

   _____

   _____

   Reference: _____

5. How is the load for a section of multioutlet assembly (plug-mold) determined?

Answer: _____

_____

_____

_____

Reference: _____

6. A motor is connected to a branch-circuit breaker in a panel that is out of sight of the motor location. The breaker is used as the controller and also the disconnecting means for the motor. If this breaker is of the "lock off" type, would this meet Code requirements?

Answer: _____

_____

_____

_____

Reference: _____

7. A No. 12 remote control circuit is protected by 80-ampere fuses in the motor disconnect (magnetic contact). Is this control circuit properly protected?

Answer: _____

_____

_____

_____

Reference: _____

8. Bus duct is reduced in size from 1,000 amperes to 400-ampere bus duct and runs 40 feet to where it terminates in a distribution panel. Is overcurrent protection required?

Answer: _____

_____

_____

_____

Reference: _____

9. Can UF cable be used from the meter, down the pole, then underground 20 feet to the disconnect for a pasture pump?

Answer: _____

_____

_____

_____

Reference: _____

10. On a 480/277-volt, 3-phase, 4-wire system where the neutral is not used, must it be grounded at the service panel?

Answer: _____

_____

_____

_____

Reference: _____

11. If available on the premises, the Code requires the bonding of all grounding electrodes together to form the grounding-electrode system. Are these requirements the same for a separately derived system?

Answer: _____

_____

_____

_____

Reference: _____

12. There is a definition in the Code of a "tap." Define a tap and is it permissible to make a tap from a tap?

Answers: _____

_____

_____

_____

Reference: _____

13. Can two tandem circuit breakers mounted adjacent to each other in a panelboard with adjacent handles tied together be used in place of two double-pole breakers to supply two 220-volt electric baseboard heaters?

Answer: _____

_____

_____

_____

Reference: _____

14. What minimum size copper grounding-electrode conductor can you use to ground a 400-ampere lighting service when two No. 3/0 copper conductors in parallel are used for the phase conductors?

Answer: _____

_____

_____

_____

Reference: _____

15. Can *aluminum* rigid metal conduit be used in hazardous locations such as in sewer plants?

Answer: _____

_____

_____

_____

Reference: _____

16. A two-lamp exit light is supplied from both a battery-operated unit equipment and a normal source. Must the normal supply to the exit light be from a recognized emergency source such as ahead of the service disconnect?

Answer: _____

_____

_____

_____

Reference: _____

17. A service is changed and moved to a different location. The old electric range run is SE cable that is too short to reach the new service so a metal junction box is installed and a short piece of new SE cable is installed to reach the new service. This is in a residential basement on a 7-foot ceiling. Does this metal junction box require grounding, and if so, does it require a separate grounding conductor to the service equipment (not the neutral of the 3-wire SE cable)?

Answer: _____

_____

_____

_____

Reference: _____

18. Can service conductors be run directly to a fire pump controller?

Answer: _____

_____

_____

_____

Reference: _____

19. A trucking firm has added a second service. They have 240-volt service to the old part of the building and 208-volt service to the new part. Common computer equipment is connected to both services with shielded TWINAX (signal cable) between devices. The two services are not tied to the same ground. This causes ground loop on signal shield. What is the best way per the *NEC®* to achieve equal potential bonding on network?

Answer: _____

_____

_____

_____

Reference: _____

20. A lighting fixture is installed over a hydromassage bathtub. Is GFCI protection required for the fixture?

Answer: _____

_____

_____

_____

Reference: _____

21. Frequently the *NEC®* requires GFCI protection of receptacles. Is there any requirement that limits the number of receptacles that can be protected by a single GFCI circuit breaker or a feed-through receptacle?

Answer: _____

_____

_____

_____

Reference: _____

22. Can an electrical discharge fixture such as a fluorescent strip be used as a branch-circuit junction box or termination point for circuit conductors?

Answer: _____

_____

_____

_____

Reference: _____

23. Does the Code allow electrical nonmetallic tubing (ENT) to be installed above a suspended ceiling that provides a thermal barrier of at least a 15-minute rating of fire-rated assemblies?

    Answer: _____

    _____

    _____

    _____

    Reference: _____

24. A $4 \times 2\frac{1}{8}$ inch square nonmetallic box is marked with the number of 14, 12, and 10 conductors that can be installed in it. Does this mean that No. 8 conductors are not permitted in the box?

    Answer: _____

    _____

    _____

    _____

    Reference: _____

25. Must lay-in fluorescent lighting fixtures be fastened or can they lay in grid T-bars of a suspended ceiling system?

    Answer: _____

    _____

    _____

    _____

    Reference: _____

PRACTICE EXAM 11

1. When UF cable is used for interior wiring as permitted, are the conductors required to be rated at 90°C?

   Answer: _____

   _____

   _____

   _____

   Reference: _____

2. Are swimming pool underwater lighting fixtures operating at less than 15 volts between conductors required to be protected by GFCI device or circuit breaker?

   Answer: _____

   _____

   _____

   _____

   Reference: _____

3. Can handle locks on circuit breakers be locked so the power to loads such as emergency lighting, sump pumps, alarm warning circuits, and other types of equipment cannot be cut off by mistake? If so, what sections of the NEC® would apply?

   Answer: _____

   _____

   _____

   _____

   Reference: _____

4. Can nonmetallic sheathed cable be run parallel to framing members through cold air returns?

   Answer: _____

   _____

   _____

   _____

   Reference: _____

5. Is a GFCI properly protected circuit supplying an outdoor portable sign an acceptable method for installation in accordance with the NEC®?

   Answer: _____

   _____

   _____

   _____

   Reference: _____

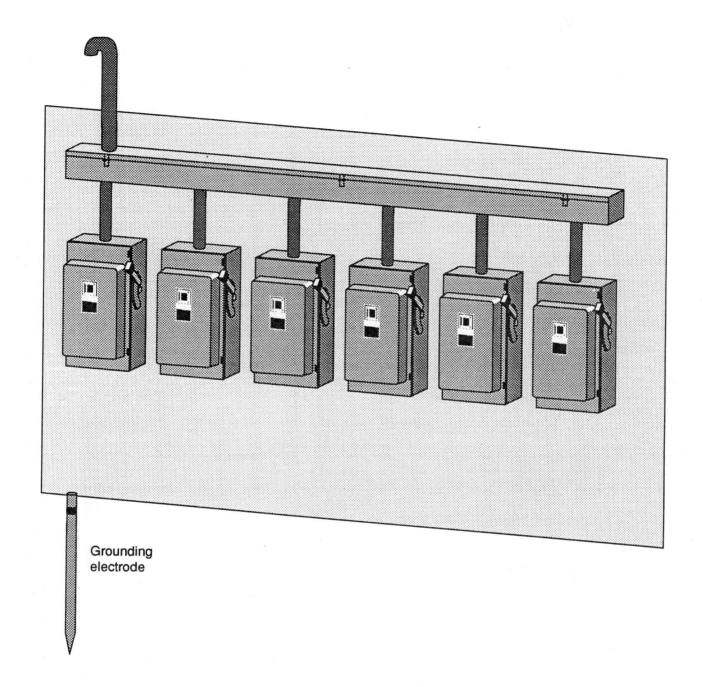

Grounding electrode

6. *Refer to the diagram above.* Where should the grounding-electrode conductor terminate in a multiple disconnect service?

Answer: _____

_____

_____

_____

Reference: _____

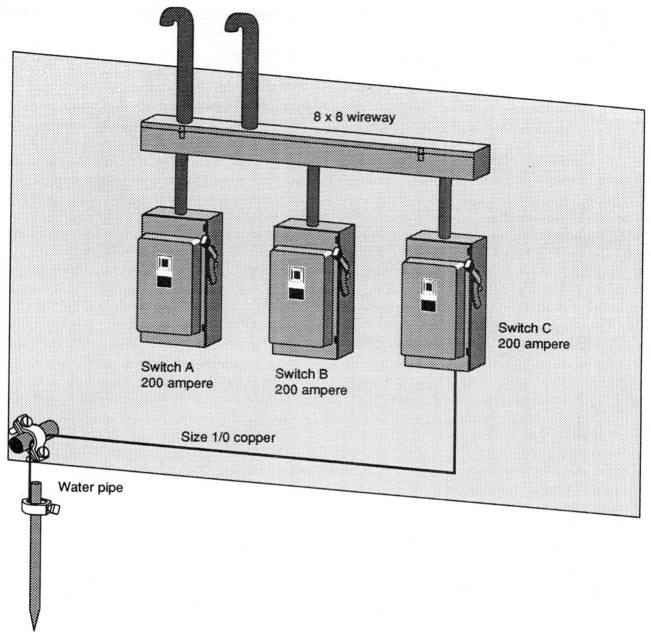

8 x 8 wireway

Switch C
200 ampere

Switch A
200 ampere

Switch B
200 ampere

Size 1/0 copper

Water pipe

Refer to the diagram above for Questions 7 and 8.

7. A 600-ampere service is built with parallel sets of 350 KCM CU THWN service conductors. Three 200-ampere fused service disconnects are supplied with No. 4/0 THWN service taps within a 8" × 8" wireway. All service bonding is continuous from the CT cabinet through all service enclosures. Does the Code permit only one grounding-electrode conductor connection at switch "C" if sized for No. 700 KCM service conductors?

Answer: _____

_____

_____

_____

Reference: _____

8. With reference to Question 7, how should the trough above the three service disconnects be sized? Is the trough to be sized as a wireway or a junction box?

Answer: _____

_____

_____

_____

Reference: _____

9. A EMT tubing and steel box system was installed throughout an office/warehouse building. An additional equipment grounding conductor was pulled with all receptacle branch-circuits and terminated at all receptacle grounding screws. Must the equipment grounding conductor also be attached to each metal box?

Answer: _____

_____

_____

_____

Reference: _____

10. Is an insulated equipment grounding conductor required in nonmetallic sheathed cable that is supplying a swimming pool pump motor installed in the garage of a single-family dwelling?

Answer: _____

_____

_____

_____

Reference: _____

11. A 225-ampere rated power panel has main lugs only and powers a calculated load of 180 amperes. It is supplied by a 500 KCM CU feeder with 400-ampere overcurrent protection. Does this installation comply with the Code?

Answer: _____

_____

_____

_____

Reference: _____

12. If a small dental office uses limited amounts of inhalation anesthetics, does Section 517-32 require standby emergency power to supply emergency lighting and the anesthetic equipment?

Answer: _____

_____

_____

_____

Reference: _____

13. Does the Code allow derating of loads in a commercial laundry? The normal load of a washing machine is 4 amperes while the spin cycle draws 9 amperes. Because it is unlikely that all machines will be spinning at the same time, what load figure should be used for load calculations?

Answer: _____

_____

_____

_____

Reference: _____

14. Is the examining room of a dental office considered to be general care patient care area? Is NM cable a permitted wiring method in this area?

Answer: _____

_____

_____

_____

Reference: _____

15. Do the metal enclosures for individual channel-type neon letters (either front-lit or back-lit) require grounding?

Answer: _____

_____

_____

_____

Reference: _____

16. Is it permissible to ground the secondary of a separately derived system back to the neutral terminal of a service switchboard instead of going directly to the water line when there is no effectively grounded structural steel nearby?

Answer: _____

_____

_____

_____

Reference: _____

17. When conductors are paralleled, does the ampacity double (assuming the same type, size, length)?

Answer: _____

_____

_____

_____

Reference: _____

18. Suppose you use three 300 Kcmil conductors in a raceway in free air with 75°C temperature rating of 285 amperes for each conductor. Instead of this, can you use six No. 1/0 conductors with 75°C rating and two conductors in parallel per phase, all in the same conduit at 300 amperes ampacity, or must you derate because you have more than three conductors? (Assume that you are using Type THWN conductors in a 2½ inch conduit with device terminations rated for 75°C.)

Answer: _____

_____

_____

_____

Reference:_____

19. Are medium base HID lampholders permitted to be installed on a 277-volt circuit?

Answer: _____

_____

_____

_____

Reference:_____

20. Is GFCI protection required for receptacles located on decks where the receptacles are located less than 6 feet 6 inches above grade and access to the deck is by steps or stepping onto a low deck? (The receptacle cannot be reached while standing on the ground.)

Answer: _____

_____

_____

_____

Reference:_____

21. Are bathroom GFCI receptacles permitted to supply lighting outlets because tripping of the GFCI causes the lights to go out?

Answer: _____

_____

_____

_____

Reference:_____

22. Is it acceptable to connect a garbage disposal and dishwasher on the same branch-circuit?

    Answer: _____

    _____

    _____

    _____

    Reference: _____

23. Are receptacles required in a four-car garage at an apartment building if the garage is detached and provided with electric power and lights?

    Answer: _____

    _____

    _____

    _____

    Reference: _____

24. Does an outdoor entrance lighting fixture with integral photocell satisfy the Code without a wall switch control? What about a motion detector used to activate the lighting fixture?

    Answer: _____

    _____

    _____

    _____

    Reference: _____

25. Does the Code require feeder conductors to have an ampacity of 125% of the continuous load plus the noncontinuous load, or is it only necessary for the feeder overcurrent device to have this rating? If only the feeder overcurrent device requires this rating, can the feeder conductors be sized only for total continuous plus noncontinuous load?

    Answer: _____

    _____

    _____

    _____

    Reference: _____

PRACTICE EXAM 12

1. In calculating feeder loads for electric space heating, is it required to figure 125% of space heating loads or is the 125% required only for branch-circuit loads?

    Answer: _____

    _____

    _____

    _____

    Reference: _____

2. For household range load calculations, is it acceptable to combine ovens and countertops and treat them as one appliance for feeder calculations?

    Answer: _____

    _____

    _____

    _____

    Reference: _____

3. Because the disconnecting means for separate buildings on one property are required to be suitable for use as service equipment, are the number of disconnects at each building's entrance limited to six?

    Answer: _____

    _____

    _____

    _____

    Reference: _____

4. The Code requires multiple services to be grounded to the same electrode. Must the grounding connections be made to same point on the electrode or can they be grounded to the same electrode (water line or structural steel) at different locations?

    Answer: _____

    _____

    _____

    _____

    Reference: _____

5. Does Article 514 apply to underground wiring (such as to a sign) that is at least 30 feet from any hazardous location in a service station?

    Answer: _____

    _____

    _____

    _____

    Reference: _____

6. What is the required minimum clearance between a thermally protected Type IC recessed incandescent fixture and wood framing?

   Answer: _____

   _____

   _____

   _____

   Reference: _____

7. Does the Code permit the metal coaxial sheath of CATV cable to be grounded to a separate driven rod or is it required to be bonded to the building's power system ground or service equipment enclosure?

   Answer: _____

   _____

   _____

   _____

   Reference: _____

8. An automotive repair garage is classified as Class I, Division 2 up 18 inches above the floor. An attached sales and office area is not classified. Is Type NM-B cable permitted to be used in the sales and office area?

   Answer: _____

   _____

   _____

   _____

   Reference: _____

9. Do Type AC and MC cables installed on 277-volt circuits require bonding if connected into concentric knockouts?

   Answer: _____

   _____

   _____

   _____

   Reference: _____

10. A swimming pool is installed at a single-family dwelling. Can 2-wire Type NM-B cable with a bare grounding conductor be used to connect the pump motor located in the basement?

   Answer: _____

   _____

   _____

   _____

   Reference: _____

11. Ceiling lighting consists of lay-in fluorescent fixtures. Is it acceptable to use ⅜ inch flexible metal conduit from fixture to fixture in lengths less than 6 feet provided an equipment grounding conductor is installed in the conduit?

    Answer: _____

    _____

    _____

    _____

    Reference: _____

12. Can a 12-volt dry-niche lighting fixture mounted on the outside of a permanent aboveground swimming pool be supplied by flexible cord connected to a receptacle located 11 feet from the pool?

    Answer: _____

    _____

    _____

    _____

    Reference: _____

13. Can a timer switch for a spa be located within 5 feet of the spa? What if it is located on the unit and not on the wall?

    Answer: _____

    _____

    _____

    _____

    Reference: _____

14. Is there any requirement relative to the location of the wall switch for an outdoor entrance light? As an example, must the switch for an entrance light at a sunroom door be located at the entrance door from outside or can it be located 10 feet away at another door leading to the sunroom?

    Answer: _____

    _____

    _____

    _____

    Reference: _____

15. When calculating the demand for a 3 kW and a 6 kW range, would you use Table 220-19, Column B for the 3 kW and Column C for the 6 kW or would you combine them and use either B or C for 2 ranges?

    Answer: _____

    _____

    _____

    _____

    Reference: _____

16. Can more than one receptacle in a laundry room be supplied by the laundry branch-circuit?

    Answer: _____

    _____

    _____

    _____

    Reference: _____

17. For a large bathroom with provisions for a washing machine, is GFCI protection required for an inaccessible receptacle located behind and dedicated to the washing machine?

    Answer: _____

    _____

    _____

    _____

    Reference: _____

18. Is there an acceptable method whereby control wiring can be run in the same conduit with power conductors to a central air-conditioning unit?

    Answer: _____

    _____

    _____

    _____

    Reference: _____

19. Does the Code require grounding the metal sheath of CATV cable entering a dwelling?

    Answer: _____

    _____

    _____

    _____

    Reference: _____

20. The Code requires bonding a metallic conduit enclosing a grounding-electrode conductor. How is the bonding jumper sized when the conduit enters a concentric knockout?

    Answer: _____

    _____

    _____

    _____

    Reference: _____

21. Does the Code require a GFCI-protected receptacle when installed within 6 feet of a dwelling wet bar or laundry sink?

    Answer: _____

    _____

    _____

    _____

    Reference: _____

22. A 500 kcmil Type THW CU feeder protected with 400-ampere overcurrent device is routed through a junction box. What is the minimum size tap conductor required to supply a 30-ampere load using the 25-foot tap rule? What is the minimum size using the 10-foot tap rule?

    Answer: _____

    _____

    _____

    _____

    Reference: _____

23. A large recreation room in a dwelling has three sets of sliding doors opening onto a covered porch. Is a switch required at each door to control porch lighting?

    Answer: _____

    _____

    _____

    _____

    Reference: _____

24. Existing feeder conduits are being repulled using conductors with 90°C Type THHN insulation. What ampacity values are allowed for these conductors that are connected to a circuit breaker or panel main lug terminals?

    Answer: _____

    _____

    _____

    Reference: _____

25. Branch-circuits serving patient care areas must be installed in metal raceways. Which areas in nursing homes and residential custodial care facilities are designated as patient care areas?

    Answer: _____

    _____

    _____

    Reference: _____

PRACTICE EXAMINATION

The following test was developed and administered by a nationally known seminar presenter and author at a meeting of Code authorities. Although this test is not representative of national electricians' tests, it does provide a good exercise in researching the *NEC®*. You will find the questions to be somewhat tricky in some cases. However, if you can pass this examination within one hour, you should be able to pass most national examinations. It will exercise your knowledge of the *NEC®*, your ability to use the document as a reference, and your ability to research variable difficult questions.

> **Each lesson is designed purposely to require the student to apply the entire *NEC®* text, not specific chapters or articles. It has been found that when studying for a timed, open-book examination the student must gain proficiency in the Table of Contents, the Index, and the ability to move quickly from cover to cover to find the correct answers to each question in a timely fashion.**

1. Equipment enclosed in a case or cabinet that is provided with a means of sealing or locking so that live parts cannot be made accessible without opening the enclosure is said to be
    A. guarded.
    B. protected.
    C. sealable equipment.
    D. lockable equipment.

    Answer: _____   Reference: _____

2. The letter(s) _____ indicate two insulated conductors laid parallel within an outer nonmetallic covering.
    A. D
    B. M
    C. T
    D. II

    Answer: _____   Reference: _____

3. Class 1 circuit conductors shall be protected against overcurrent
    A. in accordance with the values specified in Table 310-16 through 310-31 for No. 14 and larger.
    B. shall not exceed 7 amperes for No. 18.
    C. shall not exceed 10 amperes for No. 16.
    D. and derating factors do not apply.
    E. all of the above

    Answer: _____   Reference: _____

4. Enclosures for switches or circuit breakers shall not be used as
    A. junction boxes.
    B. auxiliary gutters.
    C. raceways.
    D. all of the above

    Answer: _____   Reference: _____

5. A plug fuse of the edison base has a maximum rating of _____ amperes.
    A. 20
    B. 30
    C. 40
    D. 50
    E. 60

    Answer: _____   Reference: _____

6. Power-limited fire-protective signaling circuit conductors shall be
    A. solid copper.
    B. bunch-tinned stranded copper.
    C. bonded stranded copper.
    D. any of the above

    Answer: _____   Reference: _____

7. When conduit or tubing nipples having a maximum length not to exceed _____ inches are installed between boxes and similar enclosures, the fill shall be permitted to 60 percent.
    A. 6
    B. 12
    C. 24
    D. 30

    Answer: _____   Reference: _____

8. Circuit breakers shall be
    A. capable of being opened by manual operation.
    B. capable of being closed by manual operation.
    C. trip free.
    D. all of the above
    E. A and B only

    Answer: _____   Reference: _____

9. For a circuit operating at less than 50 volts, standard lampholders shall have a rating not less than _____ watts shall be used.
    A. 300
    B. 550
    C. 660
    D. 770

    Answer: _____   Reference: _____

10. No. 18 TFF wire is rated _____ amperes.
    A. 14
    B. 10
    C. 8
    D. 6

    Answer: _____   Reference: _____

11. A No. 10 AWG solid copper wire has a cross-section area of _____ square inches.
    A. .008
    B. .101
    C. .022
    D. .160

    Answer: _____   Reference: _____

12. Conductors of light or power can occupy the same enclosure or raceway with conductors of power-limited fire-protective signaling circuits.
    A. True
    B. False

    Answer: _____   Reference: _____

13. Explanatory material in the *National Electrical Code*® is in the form of
    A. footnotes.
    B. fine print notes.
    C. obelisks and asterisk.
    D. red print.

    Answer: _____   Reference: _____

14. Straight runs of 1¼ inch rigid metal conduit using threaded couplings can be secured at not more than _____ foot intervals.
    A. 5
    B. 10
    C. 12
    D. 14

    Answer: _____   Reference: _____

15. An isolating switch is one that
    A. is not readily accessible to persons unless special means for access are used.
    B. is capable for interrupting the maximum operating overload current of a motor.
    C. is intended for use in general distribution and branch-circuits.
    D. is intended for isolating an electrical circuit from the source of power.

    Answer: _____   Reference: _____

16. Where damage to remote control circuits of safety-control equipment would produce a hazard, all conductors of this Class 1 circuit shall be installed in _____ or otherwise suitably protected from physical damage.
    I.   mineral-insulated or metal clad cable
    II. rigid metallic conduit
    III. rigid nonmetallic conduit
    A. I only
    B. II only
    C. III only
    D. I, II, or III

    Answer: _____   Reference: _____

17. _____ on equipment to be grounded shall be removed from contact surfaces to assure good electrical continuity.
    A. Conductive coatings
    B. Nonconductive coatings
    C. Manufacturers instructions
    D. all of the above

    Answer: _____   Reference: _____

18. A _____ conductor is one having one or more layers of nonconducting materials that are not recognized by this Code.
    A. noninsulating
    B. bare
    C. covered
    D. none of these

    Answer: _____   Reference: _____

19. To guard live parts, _____ accessible to qualified persons only.
    A. isolate in a room
    B. locate on a balcony
    C. enclose in a cabinet
    D. any of these

    Answer: _____   Reference: _____

20. Conductors larger than No. 4/0 are measured in
    A. inches.
    B. circular mils.
    C. square inches.
    D. AWG.

    Answer: _____   Reference: _____

21. Material identified by the superscript letter "x" includes text extracted from
    A. other *NEC*® articles and sections.
    B. other NFPA documents.
    C. other IOWA publications.
    D. none of the above

    Answer: _____   Reference: _____

22. The voltage of a circuit is defined by the Code as the _____ root-mean-square (effective) difference of potential between any two conductors in the circuit.
    A. lowest
    B. greatest
    C. average
    D. nominal

    Answer: _____   Reference: _____

23. Doorbell wiring, rated as Class 2, in a residence _____ run in the same raceway with light and power conductors.
    A. is permitted with 600-volt insulation
    B. shall not be
    C. shall be
    D. is permitted if insulation is equal to highest installed

    Answer: _____   Reference: _____

24. Equipment or materials to which has been attached a symbol or other identifying mark acceptable to the authority having jurisdiction is known as
    A. listed.
    B. labeled.
    C. approved.
    C. rated.

    Answer: _____   Reference: _____

25. The grounded conductor of a branch-circuit shall be identified by a _____ color.
    I.   natural gray
    II. continuous green
    III. continuous white
    IV. green with yellow stripe
    A. I and III
    B. II and IV
    C. III only
    D. II only

    Answer: _____   Reference: _____

26. Circuits for lighting and power shall not be connected to any system containing
    A. hazardous material.
    B. trolley wires with ground returns.
    C. poor wiring methods.
    D. dangerous chemicals or gasses.

    Answer: _____   Reference: _____

27. The minimum clearance between an electric space heating cable and an outlet box shall not be less than _____ inches.
    A. 8
    B. 12
    C. 18
    D. 6

    Answer: _____   Reference: _____

28. The ampacity requirements for a disconnecting means of x-ray equipment shall be based on _____% of the input required for the momentary rating of the equipment if greater than the long-time rating.
    A. 125
    B. 100
    C. 50
    D. none of the above

    Answer: _____   Reference: _____

29. Open conductors installed outside shall be separated from open conductors of other circuits by not less than _____ inches.
    A. 4
    B. 6
    C. 8
    D. 10
    E. 12

    Answer: _____   Reference: _____

30. The minimum headroom of working spaces about motor control centers shall be
    A. 3½ feet.
    B. 5 feet.
    C. 6 feet, 6 inches.
    D. 6 feet, 3 inches.

    Answer: _____   Reference: _____

31. Where fixed multioutlet assemblies are employed in locations where a number of appliances are likely to be used simultaneously, each foot or fraction thereof shall be considered as an outlet of not less than _____ VA.
    A. 180
    B. 200
    C. 60
    D. 35

    Answer: _____   Reference: _____

32. Soft-drawn or medium-drawn copper lead in conductors for television equipment antenna systems shall be permitted where the maximum span between points of support is less than _____ feet.
    A. 35
    B. 30
    C. 20
    D. 10

    Answer: _____   Reference: _____

33. For a one-family dwelling, the service disconnect shall not be less than _____.
    A. 60
    B. 100
    C. 150
    D. 200

    Answer: _____   Reference: _____

34. Which of the following machines shall be provided with speed limiting devices or other speed limiting means?
    I.   series motors
    II. induction
    III. self-excited DC motors
    A. I only
    B. II only
    C. III only
    D. I and III

    Answer: _____   Reference: _____

35. Where within _____ feet of any building or other structure, open wiring on insulators shall be insulated or covered.
    A. 4
    B. 3
    C. 6
    D. 10

    Answer: _____   Reference: _____

36. High-voltage conductors in tunnels shall be installed in
    A. rigid conduit.
    B. MC cable.
    C. other metal raceways.
    D. any of the above

    Answer: _____   Reference: _____

37. For garages and outbuildings on residential property, a _____ suitable for use on branch-circuits shall be permitted as the disconnecting means.
    A. snap switch
    B. set of three-way or four-way snap switches
    C. set of two-way or four-way snap switches
    D. A or B
    E. none of the above

    Answer: _____   Reference: _____

38. Utilization equipment fastened in place connected to a branch-circuit with other loads shall not exceed _____% of the branch-circuit rating.
    A. 50
    B. 60
    C. 80
    D. 100

    Answer: _____   Reference: _____

39. The maximum amperage rating of a 4 inch × ½ inch busbar is _____ amperes.
    A. 500
    B. 1000
    C. 700
    D. 2000

    Answer: _____   Reference: _____

40. For temporary wiring over 600 volts, nominal, _____ shall be provided to prevent access of other than authorized and qualified personnel.
    I. fending
    II. barriers
    III. signs
    A. I only
    B. II only
    C. III only
    D. I or II

    Answer: _____   Reference: _____

41. Each continuous-duty motor _____ horsepower or less, not permanently installed, is nonautomatically started, and is within sight of the controller shall be permitted to be protected against overload by the branch-circuit protective device.
    A. ⅛
    B. ½
    C. ¾
    D. 1

    Answer: _____   Reference: _____

42. The neutral (grounded) conductor in a mobile home shall be insulated from the equipment grounding system to which of the following location(s)?
   A. range and oven
   B. distribution panel
   C. clothes dryer
   D. all of the above

   Answer: _____   Reference: _____

43. The bottom of sign and outline lighting enclosures shall not be less than _____ feet above areas accessible to vehicles.
   A. 12
   B. 14
   C. 16
   D. 18

   Answer: _____   Reference: _____

44. The conductor used to ground the outer cover of a coaxial cable shall be
   I. insulated.
   II. No. 14 AWG minimum.
   III. guarded from physical damage when necessary.
   A. I only
   B. II only
   C. III only
   D. I, II, III

   Answer: _____   Reference: _____

45. Where energized live parts are exposed, the minimum clear work space shall not be less than _____ feet high for over 600 volts.
   A. 3
   B. 5
   C. 3½
   D. 6¼
   E. 6½

   Answer: _____   Reference: _____

46. Exposed metal raceways' surfaces shall not exceed _____ mV differences at frequencies of 1,000 Hertz or less measured across a 1,000-ohm resistance between surfaces in a general patient care area of a health care facility.
   A. 100
   B. 250
   C. 500
   D. 1,000

   Answer: _____   Reference: _____

47.   Conductors No. _____ or larger supported on solid knobs shall be securely tied thereto by tie wires having an insulation equivalent to that of the open wire.
      A. 14
      B. 12
      C. 10
      D. 8
      E. 6

      Answer: _____   Reference: _____

48.   A branch-circuit that supplies a number of outlets for lighting and appliances is known as a
      A. general purpose branch-circuit.
      B. a multi-purpose branch-circuit.
      C. a utility branch-circuit.
      D. none of the above

      Answer: _____   Reference: _____

49.   The minimum size conductor that can be used for an overhead feeder from a residence to a remote garage is No.
      A. 10 cu.
      B. 12 cu.
      C. 6 al.
      D. 10 al.

      Answer: _____   Reference: _____

50.   Transformers with a primary over 600 volts for outline lighting installations shall have secondary current ratings not more than
      A. 30 amperes.
      B. 20 amperes.
      C. 300 milliamperes.
      D. 60 amperes.

      Answer: _____   Reference: _____

FINAL EXAMINATION

Examination Instructions:

For a positive evaluation of your knowledge and preparation awareness, you must:

1. Locate yourself in a quiet atmosphere (room by yourself).
2. Have with you at least two sharp No. 2 pencils, the 1999 *NEC*®, and a hand-held calculator.
3. Time yourself (three hours) with no interruptions. Allow yourself no more than 3½ minutes per question.
4. After the test is complete, grade yourself honestly and concentrate your studies on the sections of the *NEC*® in which you missed the questions.

Caution: Do not just look up the correct answers, as the questions in this examination are only an exercise and not actual test questions. Therefore, it is important that you be able to quickly find answers from throughout the *NEC*®.

1. According to the *National Electrical Code*®, open conductors for communication equipment on a building shall be separated at least _____ feet from *lightning* conductors.
    A. 2
    B. 4
    C. 6
    D. 8

    Answer: _____   Reference: _____

2. A megohmmeter is an instrument used for
    A. polarizing a circuit.
    B. measuring high resistances.
    C. shunting a generation system.
    D. determining amperes.

    Answer: _____   Reference: _____

3. The total opposition to alternating current in a circuit that includes resistance, inductance, and capacitance is called
    A. reactance.
    B. resistance.
    C. reluctance.
    D. impedance.

    Answer: _____   Reference: _____

4. A 240-volt single-phase circuit has a resistive load of 8,500 watts. The net computed current to supply this load is _____ amps.
    A. 35
    B. 39
    C. 44
    D. 71

    Answer: _____   Reference: _____

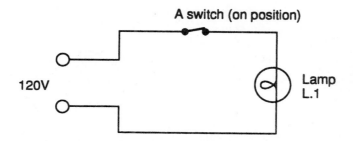

Refer to the diagram above for Questions 5 and 6.

5. In the diagram above, switch S1 is in the "on" position, but light L1 does **not** come on. Voltage across L1 is measured to be 120 volts. Voltage across S1 is measured to be 0 volt. The light does not come on because
    A. the light is open (burned out).
    B. the light and switch are shorted.
    C. the light is good but the switch does **not** make contact.
    D. there is a break in the wire of the circuit.

    Answer: _____   Reference: _____

6. In the diagram above, with a 9 amp current in the circuit, the power factor is _____ percent.
    A. 71
    B. 83
    C. 93
    D. 108

    Answer: _____   Reference: _____

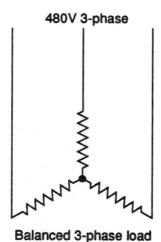

480V 3-phase

Balanced 3-phase load

7. In the diagram above, three *balanced resistive* 100-ampere loads are connected to a 480-volt, 3-phase, 3-wire circuit. The total power of this circuit is _____ kilowatts.
    A. 48
    B. 72
    C. 83
    D. 144

    Answer: _____   Reference: _____

8. To get the maximum total resistance using three resistors,
A. all three resistors should be connected in series.
B. all three resistors should be connected in parallel.
C. two resistors should be connected in parallel then one connected in series.
D. two resistors should be connected in series then one connected in parallel.

Answer: _____   Reference: _____

9. Four resistance heaters are connected in parallel. Their resistances are heater 1, 20 ohms; heater 2, 30 ohms; heater 3, 60 ohms; and heater 4, 10 ohms. The total resistance of the parallel circuit is _____ ohms.
A. 5
B. 12
C. 50
D. 120

Answer: _____   Reference: _____

10. A building on a blueprint is 16 inches × 10 inches. If the drawing scale is ¼ inch = 1 foot, what is the area of the building in square feet?
A. 160 square feet
B. 640 square feet
C. 2,560 square feet
D. 5,120 square feet

Answer: _____   Reference: _____

The following diagram is for Question 11.

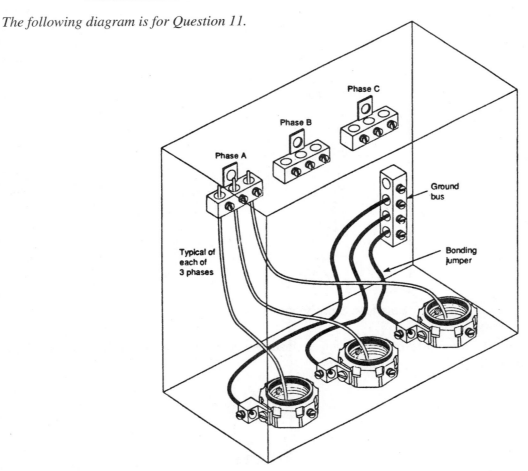

11.   In the diagram on page 234, each of the supply side EMT conduits contain three 500 MCM copper THW service conductors in parallel. As shown, a separate bonding jumper is installed from each conduit to the grounded bus terminal.

Each bonding jumper must be at least size _____ AWG copper.
A.   1/0
B.   2/0
C.   3/0
D.   4/0

Answer: _____   Reference: _____

12.   A single-phase, 3-wire service has two ungrounded conductors of size 2/0 AWG copper THWN. The neutral conductor is size 1 AWG copper THWN. The conduit size needed for the service-entrance conductors is at least size _____ inch.
A.   1½
B.   2
C.   2½
D.   3

Answer: _____   Reference: _____

13.   A metallic cold water piping system is available within the structure being supplied by an electrical service with size 1/0 THW copper service-entrance conductors. The copper grounding-electrode conductors run to the metal water pipe shall have a minimum size of:
A.   No. 8 AWG
B.   No. 6 AWG
C.   No. 4 AWG
D.   No. 2 AWG

Answer: _____   Reference: _____

14.   Ground-fault protection is required to be installed on a 2,000 amp, solidly grounded, wye service. What is the maximum setting of the ground fault protection?
A.   800 amperes
B.   1,000 amperes
C.   1,200 amperes
D.   1,600 amperes

Answer: _____   Reference: _____

15.   Overhead service conductors shall be installed so that the minimum clearance from a window opening is
A.   two feet.
B.   three feet.
C.   six feet.
D.   eight feet.

Answer: _____   Reference: _____

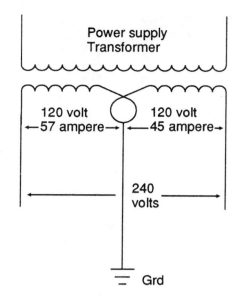

16. The 240-volt, single-phase feeder shown in the diagram above supplies a branch-circuit panel that has an improperly balanced load of 57 amperes on one ungrounded conductor and 45 amperes on the other ungrounded conductor. What is the load on the grounded (neutral) conductor?
    A. No load
    B. 12 amperes
    C. 57 amperes
    D. 102 amperes

    Answer: _____   Reference: _____

17. The maximum number of power or lighting conductors that can be installed in a raceway before the derating factors must be applied is
    A. 1.
    B. 3.
    C. 5.
    D. 7.

    Answer: _____   Reference: _____

18. A 230-volt, single-phase, 100-ampere circuit is installed in a nonmetallic raceway. Which of the following conditions must apply to the equipment grounding conductor installed with the circuit?
    I. It must be counted when determining conductor fill of the raceway.
    II. It must be the same size as that of the circuit conductor.
    A. I only
    B. II only
    C. Both I and II
    D. Neither I nor II

    Answer: _____   Reference: _____

19. Three No. 4 THWN and four No. 1/0 THW conductors are to be installed in a single run of IMC conduit. The minimum size of conduit permitted is
    A. 1 inch.
    B. 2 inches.
    C. 3 inches.
    D. 4 inches.

    Answer: _____   Reference: _____

20. Eight No. 4 THWN conductors are to be installed in a single run of nonmetallic Schedule 40 conduit. The minimum size of conduit permitted is
    A. 1 inch.
    B. 1½ inches.
    C. 2 inches.
    D. 2½ inches.

    Answer: _____   Reference: _____

21. A service disconnecting means can be installed at which of the following locations?
    I.   Outside a building at a readily accessible point nearest the point of entrance of the service-entrance conductors.
    II. Inside a building at a readily accessible point nearest the point of entrance of the service-entrance conductors.
    A. I only
    B. II only
    C. Either I or II
    D. Neither I nor II

    Answer: _____   Reference: _____

Conductors crossing and connecting

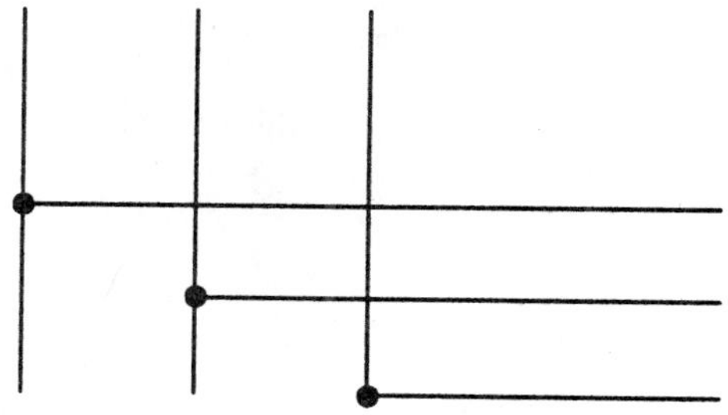

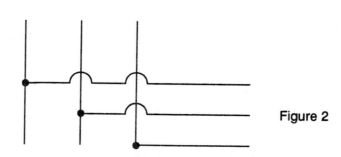

Figure 1

Figure 2

22. Figures 1 and 2 shown above are standard electrical symbols used on blueprints representing crossing conductors. Which symbol represents two conductors crossing but not connecting?
A. Figure 1 only
B. Figure 2 only
C. Figures 1 and 2
D. Neither Figure 1 nor Figure 2

Answer: _____   Reference: _____

23. The following figure is a standard symbol used on blueprints. The symbol represents which of the following?
A. An air circuit breaker
B. A lightning arrestor
C. A fuse
D. A thermal element

Answer: _____   Reference: _____

24. A motor is protected against short-circuit and ground-fault by an adjustable instantaneous trip circuit breaker that is part of a combination controller having motor overload and short-circuit and ground-fault protection in each conductor. The setting of the instantaneous trip breaker shall be permitted to exceed the motor full-load current by not more than
A. 250 percent.
B. 1,300 percent.
C. 700 percent.
D. 1,000 percent.

Answer: _____   Reference: _____

25. Thermal overload relays are used for the protection of polyphase induction motors. Their primary purpose is to protect the motor in case of
A. reversal of phases in the supply.
B. low line voltage.
C. short-circuit between phases.
D. sustained overload.

Answer: _____   Reference: _____

26. A motor controller and motor branch-circuit disconnecting means a 2,300-volt motor shall have a continuous ampere rating of
    A. not less than the trip setting of the short-circuit protective device rating.
    B. not less than the trip setting of the overload protection device.
    C. not less than the trip setting of the fault-current protection device.
    D. not less than the locked-rotor rating of the motor.

    Answer: _____   Reference: _____

27. The disconnecting means for both a motor and the controller shall be
    A. permitted within the same enclosure as the controller.
    B. located separately from the controller enclosure.
    C. located at the service equipment.
    D. permitted within the watt-hour meter enclosure.

    Answer: _____   Reference: _____

28. Speed limiting devices shall be provided with which of the following?
    A. polyphase squirrel-cage motors
    B. synchronous motors
    C. compound motors
    D. series motors

    Answer: _____   Reference: _____

29. Which of the following appliances can be grounded to the grounded (neutral) conductor?
    A. electric water heater
    B. kitchen disposal
    C. dishwasher
    D. electric dryer

    Answer: _____   Reference: _____

30. A metal lighting fixture shall be grounded if located
    A. 10 feet vertically or 6 feet horizontally from a kitchen sink.
    B. 8 feet vertically or 5 feet horizontally from a kitchen sink.
    C. 6 feet vertically or 3 feet horizontally from a kitchen sink.
    D. 8 feet vertically or 3 feet horizontally from a kitchen sink.

    Answer: _____   Reference: _____

31. When used, a driven ground rod shall be installed so the soil will be in contact with a length of the rod not less than
    A. 4 feet.
    B. 6 feet.
    C. 8 feet.
    D. 10 feet.

    Answer: _____   Reference: _____

32. Which of the following most accurately describes the condition of a motor known as "locked-rotor?"
    A. When the electrician places a lock on the motor controller to keep the motor from being energized
    B. Mechanical brakes used on the motor shaft to stop the motor during shutdown
    C. An electronic control device used to lock the speed of the motor at that specified by the manufacturer
    D. When the circuits of a motor are energized but the rotor is not turning

    Answer: _____ Reference: _____

33. The following diagram represents overhead conductors between two buildings on an industrial site. The voltage is 240/480 volts AC. The conductors pass over a driveway leading to a loading dock at one of the buildings. What is the minimum vertical clearance permitted between the overhead conductors and the driveway?
    A. 10 feet
    B. 12 feet
    C. 15 feet
    D. 18 feet

    Answer: _____ Reference: _____

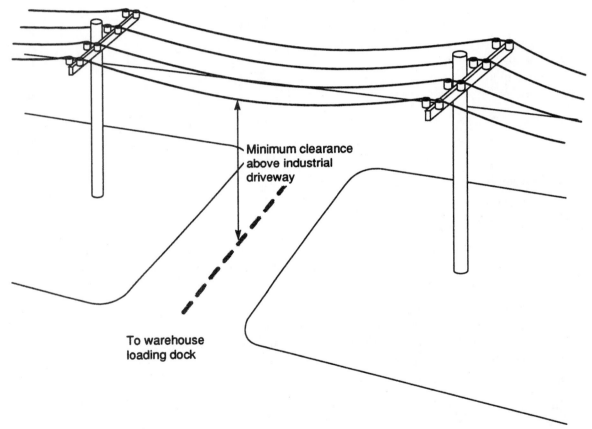

Minimum clearance above industrial driveway

To warehouse loading dock

34. When a ground-fault protection for equipment is installed within service equipment, it shall be performance tested
    A. at the factory before shipment.
    B. before being installed on site.
    C. when first installed on site.
    D. after the electrical system has been used for one week.

    Answer: _____ Reference: _____

35. Three continuous-duty motors with full-load current ratings of 5.6 amperes, 4.5 amperes, and 4.5 amperes, respectively, are to be installed on a single branch-circuit. The circuit conductors shall have a minimum ampacity of
   A. 16 amperes.
   B. 20 amperes.
   C. 24 amperes.
   D. 30 amperes.

   Answer: _____   Reference: _____

36. The following diagram represents an electrical service with a fused switch as the main disconnect. What is the maximum height the center of the grip of the operating handle of the switch is permitted to be located above the ground?
   A. 5 feet
   B. 6½ feet
   C. 7 feet
   D. ½ foot

   Answer: _____   Reference: _____

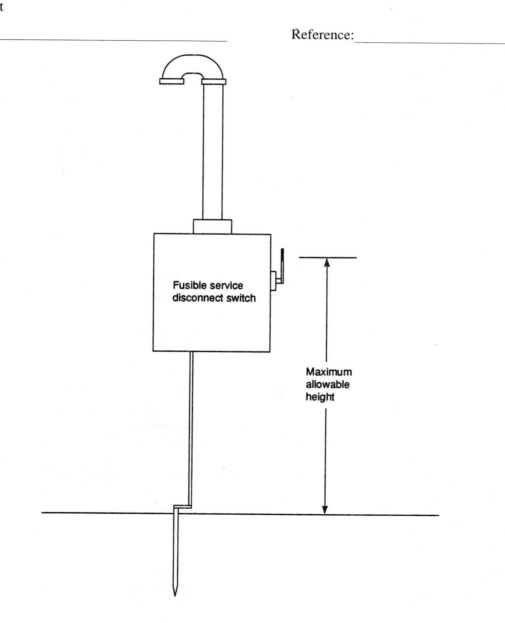

Fusible service disconnect switch

Maximum allowable height

37. How is the size of an electrical conductor supplying a circuit determined?
    A. Voltage
    B. Amperage
    C. The length
    D. all of the above

    Answer: _____   Reference: _____

38. Ohm's law is
    A. measurement of the I^2R losses.
    B. the relationship between voltage, current, and power.
    C. an equation for determining power.
    D. a relationship between voltage, current, and resistance.

    Answer: _____   Reference: _____

39. Electrical pressure is measured in
    A. volts.
    B. amperes.
    C. coulombs.
    D. watts.

    Answer: _____   Reference: _____

40. The resistance of a 1,500-watt, 120-volt resistance heater element is
    A. 14.4 ohms.
    B. 9.6 ohms.
    C. 11.2 ohms.
    D. 12.5 ohms.

    Answer: _____   Reference: _____

41. Total ampacity of stranded or solid conductors is the same if they have the same
    A. diameter.
    B. circumference.
    C. cross-sectional area.
    D. insulation.

    Answer: _____   Reference: _____

42. A branch-circuit that supplies a number of outlets for lighting and appliances is called a _____ branch-circuit.
    A. general purpose
    B. utility
    C. multipurpose
    D. none of the above

    Answer: _____   Reference: _____

43. A 10-ohm resistance carrying 10 amperes of current uses _____ watts of power.
   A. 100
   B. 200
   C. 500
   D. 1,000

   Answer: _____   Reference: _____

44. A _____ stores energy in much the same manner as a spring stores mechanical energy.
   A. coil
   B. capacitor
   C. resistor
   D. none of the above

   Answer: _____   Reference: _____

45. _____ means that it is constructed or protected so exposure to the weather will not interfere with its successful operation.
   A. Weatherproof
   B. Weather tight
   C. Weather resistant
   D. All of the above

   Answer: _____   Reference: _____

46. The conductor clearance from windows can be satisfactory if it is
   A. not less than 24 inches from the window.
   B. run above the top level of the window.
   C. run below the bottom level of the window.
   D. any of the above.

   Answer: _____   Reference: _____

47. The *NEC*® states that electrical equipment shall be installed
   A. not exceeding the provisions of the Code.
   B. not less than the Code permits.
   C. according to the Code and local code amendments.
   D. none of the above

   Answer: _____   Reference: _____

48. The equivalent resistance of three resistors of 8 ohms, 8 ohms, and 4 ohms that are connected in parallel is
   A. 2 ohms.
   B. 4 ohms.
   C. 6 ohms.
   D. 8 ohms.

   Answer: _____   Reference: _____

49. Because fuses are rated by both amperage and voltage, a fuse will operate correctly on
    A. AC only.
    B. AC or DC.
    C. DC only.
    D. any voltage.

    Answer: _____   Reference: _____

50. The following 480-volt, 3-phase, 3-wire, intermittent use equipment is in a commercial kitchen: two 5,000-watt water heaters; four 3,000-watt fryers; and two 6,000-watt ovens. Each ungrounded conductor in the feeder circuit for this kitchen equipment must be sized to carry a minimum computed load of _____ amps.
    A. 15
    B. 27
    C. 41
    D. 46

    Answer: _____   Reference: _____

Appendix 1: Symbols

(Courtesy of American Iron and Steel Institute)

Electrical Wiring Symbols
Selected from American National Standard Graphic for
Electrical Wiring and Layout Diagrams Used in Architecture and Building Construction
ANSI Y32.9-1972

1. Lighting Outlets

Ceiling   *Wall*

1.1 Surface or pendant incandescent, mercury-vapor, or similar lamp fixture

1.2 Recessed incandescent, mercury-vapor, or similar lamp fixture

1.3 Surface of pendant individual fluorescent fixture

1.4 Recessed individual fluorescent fixture

1.5 Surface or pendant continuous-row fluorescent fixture

1.6 Recessed continuous-row fluorescent fixture

1.8 Surface or pendant exit light

1.9 Recessed exit light

1.10 Blanked outlet

1.11 Junction box

1.12 Outlet controlled by low-voltage switching when relay is installed in outlet box

2. Receptacle Outlets

Grounded   *Ungrounded*

2.1 Single receptacle outlet

2.2 Duplex receptacle outlet

2.3 Triplex receptacle outlet

2.4 Quadrex receptacle outlet

2.5 Duplex receptacle outlet – split wired

2.6 Triplex receptacle outlet – split wired

2.7 Single special-purpose receptacle outlet – split wired

NOTE 2.7A: Use numeral or letter as a subscript alongside the symbol, keyed to explanation in the drawing list of symbols, to indicate type of receptacle or use

2.8 Duplex special-purpose receptacle outlet
See note 2.7A

2.9 Range outlet (typical)
  See note 2.7A

2.10 Special-purpose connection or provision for connection
  Use subscript letters to indicate function (SW – dishwasher; CD – clothes dryer, etc).

DW   UNG
                      DW

2.12 Clock hanger receptacle

© UNG

2.13 Fan hanger receptacle

Ⓕ UNG

2.14 Floor single receptacle outlet

UNG

2.15 Floor duplex receptacle outlet

UNG

2.16 Floor special-purpose outlet
  See note 2.7A

UNG

2.17 Floor telephone outlet – public

2.18 Floor telephone outlet – private

2.19 Underfloor duct and junction box for triple, double, or single duct system (as indicated by the number of parallel lines)

2.20 Cellular floor header duct

3. Switch Outlets

3.1 Single-pole switch

S

3.2 Double-pole switch

S2

3.3 Three-way switch

S3

3.4 Four-way switch

S4

3.5 Key-operated switch

SK

3.6 Switch and pilot lamp

SP

3.7 Switch for low-voltage switching system

SL

3.8 Master switch for low-voltage switching system

SLM

3.9 Switch and single receptacle

S

3.10 Switch and double receptacle

S

3.11 Door switch

SD

3.12 Time switch

ST

3.13 Circuit breaker switch

SCB

3.14 Momentary contact switch or pushbutton
for other than signaling system

SMC

3.15 Ceiling pull switch

(S)

5.13 Radio outlet

[R]

5.14 Television outlet

[TV]

6. Panelboards, switchboards, and related equipment

6.1 Flush-mounted panelboard and cabinet
  NOTE 6.1A: Identify by notation or schedule

6.2 Surface-mounted panelboard and cabinet
  See note 6.1A

6.3 Switchboard, power control center, unit
substations (should be drawn to scale)
  See note 6.1A

6.4 Flush-mounted terminal cabinet
  See note 6.1A

  NOTE 6.4A: In small-scale drawings the TC may be
indicated alongside the symbol

TC

6.5 Surface-mounted terminal cabinet
  See note 6.1A and 6.4A

TC

6.6 Pull box
  Identify in relation to wiring system section and size

6.7 Motor or other power controller
  See note 6.1A

MC

6.8 Externally operated disconnection switch
  See note 6.1A

6.9 Combination controller and disconnection means
  See note 6.1A

7. Bus Ducts and Wireways

7.1 Trolley duct
  See note 6.1A

T		T		T

7.2 Busway (service, feeder, or plug-in)
  See note 6.1A

B		B		B

7.3 Cable through, ladder, or channel
  See note 6.1A

BP		BP		BP

7.4 Wireway
  See note 6.1A

W		W		W

9. Circuiting

Wiring method identification by notation on
drawing or in specifications.

9.1 Wiring concealed in ceiling or wall

NOTE 9.1A: Use heavy weight line to identify service
and feed runs.

9.2 Wiring concealed in floor
  See note 9.1A

Appendix 2: Basic Electrical Formulas

Basic Electrical Formulas
(Courtesy of American Iron and Steel Institute)

DC Circuit Characteristics

Ohm's Law:

$$E = IR \quad I = \frac{E}{R} \quad R = \frac{E}{I}$$

E = voltage impressed on circuit (volts)

I = current flowing in circuit (amperes)

R = circuit resistance (ohms)

Resistances in Series:

$R_t = R_1 + R_2 + R_3 + \ldots$

R_T = total resistance (ohms)

R_1, R_2 etc. = individual resistances (ohms)

Resistances in Parallel:

$$R_t = \frac{1}{\frac{1}{R_1} + \frac{1}{R_2} + \frac{1}{R_3} + \ldots}$$

Formulas for the conversion of electrical and mechanical power:

$$HP = \frac{watts}{746} (watts \times .00134)$$

$$HP = \frac{kilowatts}{.746} (kilowatts \times 1.34)$$

Kilowatts = HP × .746

Watts = HP × 746

HP = Horsepower

In direct-current circuits, electrical power is equal to the product of the voltage and current:

$$P = EI = I_2R = \frac{E_2}{R}$$

P = power (watts)

E = voltage (volts)

I = current (amperes)

R = resistance (ohms)

Solving the basic formula for I, E, and R gives

$$I = \frac{P}{E} = \sqrt{\frac{P}{R}}; \quad E = \frac{P}{I} = \sqrt{RP}; \quad R = \frac{E2}{P} = \frac{P}{T^2}$$

Energy

Energy is the capacity for doing work. Electrical energy is expressed in kilowatt-hours (kWhr), one kilowatt-hour representing the energy expended by a power source of 1 kW over a period of 1 hour.

Efficiency

Efficiency of a machine, motor or other device is the ratio of the energy output (useful energy delivered by the machine) to the energy input (energy delivered to the machine), usually expressed as a percentage:

$$Efficiency = \frac{output}{input} \times 100\%$$

$$or \ Output = Input \times \frac{efficiency}{100\%}$$

Torque

Torque may be described as a force tending to cause a body to rotate. It is expressed in pound-feet or pounds of force acting at a certain radius:

Torque (pound-feet) = force tending to produce rotation (pounds) × distance from center of rotation to point at which force is applied (feet).

Relations between torque and horsepower:

$$Torque = \frac{33,000 \times HP}{6.28 \times rpm}$$

$$HP = \frac{6.28 \times rpm \ time \ torque}{33,000}$$

rpm = speed of rotating part (revolutions per minute)

AC Circuit Characteristics

The instantaneous values of an alternating current or voltage vary from zero to maximum value each half cycle. In the practical formula that follows, the "effective value" of current and voltage is used, defined as follows:

Effective value = 0.707 × maximum instantaneous value

Inductances in Series and Parallel:

The resulting circuit inductance of several inductances in series or parallel is determined exactly as the sum of resistances in series or parallel as described under dc circuit characteristics.

Impedance:

Impedance is the total opposition to the flow of alternating current. It is a function of resistance, capacitive reactance and inductive reactance. The following formulae relate these circuit properties:

$$X_L = 2\pi Hz L \quad Xc = \frac{1}{2\pi Hz C} \quad Z = \sqrt{R^2 + (X_L - X_C^2)}$$

X_L = inductive reactance (ohms)

X_c = capacitive reactance (ohms)

Z = impedance (ohms)

Hz - (Hertz) cycles per second

C - capacitance (farads)

L - inductance (henrys)
R = resistance (ohms)
π = 3.14

In circuits where one or more of the properties L, C, or R is absent, the impedance formula is simplified as follows:

Resistance only: Inductance only: Capacitance only:
$Z = R$ $Z = X_L$ $Z = X_C$
Resistance and   Resistance and   Inductance and
Inductance only: Capacitance only: Capacitance only:
$Z = \sqrt{R^2 + X_L^2}$ $Z = \sqrt{R^2 + X_C^2}$ $Z = \sqrt{X_L - X_C^2}$

Ohm's law for AC circuits:

$E = 1 \times Z$ $I = \dfrac{E}{Z}$ $Z = \dfrac{E}{I}$

Capacitances in Parallel:

$C_t = C_1 + C_2\ C_3 + ...$

C_t = total capacitance (farads)

$C_1 C_2 C_3 ...$ = individual cpacitances (farads)

Capacitances in series:

$C_t = \dfrac{1}{\dfrac{1}{C_1} + \dfrac{1}{C_2} + \dfrac{1}{C_3} + ...}$

Phase Angle

An alternating current through an inductance lags the voltage across the inductance by an angle computed as follows:

Tangent of angle of lag = $\dfrac{X_L}{R}$

An alternating current through a capacitance leads the voltage across the capacitance by an angle computed as follows:

Tangent of angle of lead = $\dfrac{X_C}{R}$

The resultant angle by which a current leads or lags the voltage in an entire circuit is called the phase angle and is computed as follows:

Cosine of phase angle = $\dfrac{R \text{ of circuit}}{Z \text{ of circuit}}$

Power Factor

Power factor of a circuit or system is the ratio of actual power (watts) to apparent power (volt-amperes), and is equal to the cosine of the phase angle of the circuit:

$PF = \dfrac{\text{actual power}}{\text{apparent power}} = \dfrac{\text{watts}}{\text{volts} \times \text{amperes}} = \dfrac{kW}{kVA} = \dfrac{R}{Z}$

KW = kilowatts

kVA = kilowatt-amperes = volt-amperes + 1,000

PF = power factor (expressed as decimal or percent)

Single-Phase Circuits

$kVA = \dfrac{EI}{1,000} = \dfrac{kW}{PF}kW = kVA \times PF$

$I = \dfrac{P}{E \times PF}E = \dfrac{P}{I \times PF}PF = \dfrac{P}{E \times I}$

$P = E \times I \times PF$

P = power (watts)

Two-Phase Circuits

$I = \dfrac{P}{2 \times E \times PF}E = \dfrac{P}{2 \times I \times PF}PF = \dfrac{P}{E \times I}$

$kVA = \dfrac{2 \times E \times I}{1000} = \dfrac{kW}{PF}kW = kVA \times PF$

$P = 2 \times E \times 1 \times PF$

E = phase voltage (volts

Three-Phase Circuits, Balanced Star or Wye

$I_N = O\ I = I_P E = \sqrt{3}E_P = 1.73E_P$

$E_P = \dfrac{E}{\sqrt{3}} = \dfrac{E}{1.73} = 0.577E$

I_N = current in neutral (amperes)

I = line current per phase (amperes)

I_P = current in each phase winding (amperes)

E = voltage, phase to phase (volts)

E_P = voltage, phase to neutral (volts)

Three-Phase Circuits, Balanced Delta

$I = 1.732 \times I_P I_P = \dfrac{1}{\sqrt{3}} = 0.577 \times I$

$E = E_P$

Power:

Balanced 3-Wire, 3-Phase Circuit, Delta or Wye

For unit power factor (PF = 1.0):

$P = 1.732 \times E \times I$

$I = \dfrac{P}{\sqrt{3}}E = 0.577\dfrac{P}{E}$ $E = \dfrac{P}{\sqrt{3}} \times I = 0.577\dfrac{P}{I}$

P = total power (watts)

For any load:

$P = 1.732 \times E \times I \times PF VA = 1.732 \times E \times I$

$E = \dfrac{P}{PF \times 1.73 \times I} = 0.577 \times \dfrac{P}{PF \times I}$

$I = \dfrac{P}{PF \times 1.73 \times E} = 0.577 \times \dfrac{P}{I} \times E$

$PF = \dfrac{P}{1.73 \times I \times E} = \dfrac{0.577 \times P}{I \times E}$

VA = apparent power (volt-amperes)

P = actual power (watts)

E = line voltage (volts)

I = line current (amperes)

Power Loss:

Any AC or DC Circuit

$P = I_2 R I = \sqrt{\dfrac{P}{R}}R = \dfrac{P}{I^2}$

P = power heat loss in circuit (watts)

I = effective current in conductor (amperes)

R = conductor resistance (ohms)

Load Calculations

Branch Circuits—Lighting & Appliance 2-Wire:

$$I = \frac{\text{total connected load (watts)}}{\text{line voltage (volts)}}$$

I = current load on conductor (amperes)

3-Wire:

Apply same formula as for 2–wire branch circuit, considering each line to neutral separately. Use line-to-neutral voltage; result gives current in line conductors.

USEFUL FORMULAS

TO FIND	SINGLE PHASE	THREE PHASE	DIRECT CURRENT
AMPERES when kVA is known	$\dfrac{kVA \times 1000}{E}$	$\dfrac{kVA \times 1000}{E \times 1.73}$	not applicable
AMPERES when horsepower is known	$\dfrac{HP \times 746}{E \times \% \text{ eff.} \times pf}$	$\dfrac{HP \times 746}{E \times 1.73 \times \% \text{ eff.} \times pf}$	$\dfrac{HP \times 746}{E \times \% \text{ eff.}}$
AMPERES when kilowatts are known	$\dfrac{kW \times 1000}{E \times pf}$	$\dfrac{kW \times 1000}{E \times 1.73 \times pf}$	$\dfrac{kW \times 1000}{E}$
HORSEPOWER	$\dfrac{I \times E \times \% \text{ eff.} \times pf}{746}$	$\dfrac{I \times E \; 1.73 \times \% \text{ eff.} \times pf}{746}$	$\dfrac{I \times E \times \% \text{ eff.}}{746}$
KILOVOLT AMPERES	$\dfrac{I \times E}{1000}$	$\dfrac{I \times E \times 1.73}{1000}$	not applicable
KILOWATTS	$\dfrac{I \times E \times pf}{1000}$	$\dfrac{I \times E \times 1.73 \times pf}{1000}$	$\dfrac{I \times E}{1000}$
WATTS	$E \times I \times pf$	$E \times I \times 1.73 \times pf$	$E \times I$

$$\text{ENERGY EFFICIENCY} = \frac{\text{Load Horsepower} \times 746}{\text{Load Input kVA} \times 1000}$$

$$\text{POWER FACTOR (pf)} = \frac{\text{Power Consumed}}{\text{Apparent Power}} = \frac{W}{VA} = \frac{kW}{kVA} = \cos\varnothing$$

I = Amperes	E = Volts	kW = Kilowatts	kVA = Kilovolt-amperes
HP = Horsepower	% eff. = Percent Efficiency e.g., 90% eff. is 0.90		pf = Power Factor e.g., 95% pf is 0.95

EQUATIONS BASED ON OHM'S LAW:

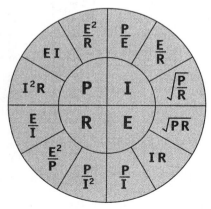

P = POWER, IN WATTS
I = CURRENT, IN AMPERES
R = RESISTANCE, IN OHMS
E = ELECTROMOTIVE FORCE, IN VOLTS

Appendix 3: 1999 Highlighted Code Changes

There were 4,259 proposals to change the 1999 *NEC*® and 3,246 public comments. The *NEC*® Technical Correlating Committee usability task group made several recommendations to make the Code more user friendly. These recommendations resulted in significant changes.

- Articles 210 and 220 were revised, relocating some branch-circuit requirements from Article 220 to 210.
- Article 250 Grounding was completely reformatted and renumbered. A cross reference was added in a new Appendix E. This cross reference is two-fold: it lists the 1996 numbers to the 1999 numbers and vice-versa.
- The notes to the 0- to 2,000-volt tables in Article 310 were rewritten into Section 310-15.
- The footnotes to the 0- to 2,000-volt tables were properly placed in Article 240 Overcurrent Protection.

- Many of the exceptions and fine print notes (FPN) were deleted or rewritten into positive language throughout the Code.
- Article 710 was deleted and the requirements for over-600-volt circuits and equipment were placed in Articles 110, 240, 300, and new Article 490.

A vertical line in the margin of the 1999 *NEC*® shows the changes and additions to the Code. The deletions are shown as a "bullet" or large black dot in the margin. However, where a section has been relocated there is nothing unique to indicate this and can cause some confusion, because it may lead the experienced user to believe a section has been deleted when, in fact, it has been moved.

The following quizzes only include questions to highlight a few of the important changes. For a complete analysis of the 1999 *NEC*® changes, see "Illustrated Changes in the 1999 *National Electrical Code*®" by Ron O'Riley.

1999 CODE CHANGE QUIZ ONE

1. Can the electric "service" supplying a building or structure be supplied from a private generator?

   Answer: _____

   _____

   _____

   _____

   Reference: _____

2. Where are the requirements for tunnel installations over 600 volts covered?

   Answer: _____

   _____

   _____

   _____

   Reference: _____

3. A dedicated space equal to the depth and width of the equipment from the floor to a height of _____ or to the structural ceiling whichever is _____.

    Answer: _____

    _____

    _____

    _____

    Reference: _____

4. Grounding and bonding requirements for tunnel installations over 600 volts are covered in Section _____.

    Answer: _____

    _____

    _____

    _____

    Reference: _____

5. Generally, grounded conductors No. 6 or smaller shall be identified by a continuous white or natural gray finish or by _____ on other than green insulation along its entire length.

    Answer: _____

    _____

    _____

    _____

    Reference: _____

6. Is a receptacle installed in an accessory building such as a storage building at grade level associated with a dwelling required to have a ground-fault circuit interrupter protection for personnel?

    Answer: _____

    _____

    _____

    _____

    Reference: _____

7. What is an arc-fault circuit interrupter?

    Answer: _____

    _____

    _____

    _____

    Reference: _____

8. May the service conductors originate at a generator where there is no utility serving the premises?

   Answer: _____

   _____

   _____

   _____

   Reference: _____

9. Other electrical equipment associated with the electrical installation located above or below shall not be allowed to extend beyond the front of electrical equipment. (True or False)

   Answer: _____

   _____

   _____

   _____

   Reference: _____

10. The requirements for dedicated electrical space were located in Article 384 in the 1996 *NEC®*. Have these requirements been deleted or moved?

    Answer: _____

    _____

    _____

    _____

    Reference: _____

11. General requirements for over 600 volts nominal were formerly located in Article 710. Have these requirements been deleted or moved?

    Answer: _____

    _____

    _____

    _____

    Reference: _____

12. An insulated grounded conductor of No. 6 or smaller shall be identified by a continuous white or natural gray (not gray color) or _____.

    Answer: _____

    _____

    _____

    _____

    Reference: _____

13. Receptacles located in garages or accessory buildings having a floor located at or below grade and are limited to storage areas, work areas, or areas of similar use are required to be of the _____ type.

Answer: _____

_____

_____

_____

Reference: _____

14. All branch-circuits supplying 125-volt, single-phase, 15- and 20-ampere receptacles installed in bedrooms shall be protected by arc-fault circuit interrupters after the date of _____.

Answer: _____

_____

_____

_____

Reference: _____

15. Are the required two small-appliance circuits permitted to serve more than one kitchen?

Answer: _____

_____

_____

_____

Reference: _____

16. In dwelling units, at least one receptacle shall be located on the wall _____ to each basin in bathrooms not more than _____ inches of the outside edge of each basin.

Answer: _____

_____

_____

_____

Reference: _____

17. In guest rooms of hotels, motels, and similar occupancies, at least _____ of the receptacles required by Section 210-52(a) shall be _____.

Answer: _____

_____

_____

_____

Reference: _____

18. An additional load of _____ shall be included for each 2 feet of track lighting or fraction thereof in all occupancies except dwelling units or guest rooms of hotels and motels.

   Answer: _____

   _____

   _____

   _____

   Reference: _____

19. Outside feeders and branch-circuits not exceeding 300 volts to ground over residential property and driveways, and commercial areas not subject to truck traffic shall have a minimum clearance from ground of _____.

   Answer: _____

   _____

   _____

   _____

   Reference: _____

20. For a one-family dwelling, the feeder disconnecting means shall have a rating of not less than _____, 3-wire.

   Answer: _____

   _____

   _____

   _____

   Reference: _____

1999 CODE CHANGE QUIZ TWO

1. Which section of the Code permits one additional set of service-entrance conductors to supply a separate meter for public or common area loads not permitted to be supplied from an individual dwelling unit?

   Answer: _____

   _____

   _____

   _____

   Reference: _____

2. For a one-family dwelling the service disconnecting means shall have a rating of not less than _____, 3-wire.

   Answer: _____

   _____

   _____

   _____

   Reference: _____

3. Do the overcurrent protection requirements in Section 240-3(d) for small conductors apply to motor or motor-operated appliances?

   Answer: _____

   _____

   _____

   _____

   Reference: _____

4. Tap conductor is defined in Section _____.

   Answer: _____

   _____

   _____

   _____

   Reference: _____

5. Where a tap is located outdoors except for the point of termination, the conductors are protected from physical damage and terminate in a single overcurrent device, the disconnecting means is readily accessible either outside or inside the building nearest the point of entrance, and the overcurrent device is integral or immediately adjacent to the disconnect, the length shall not exceed _____ feet.

   Answer: _____

   _____

   _____

   _____

   Reference: _____

6.   Where in the Code is a "Supervised Industrial Installation" defined?

     Answer: _____

     _____

     _____

     _____

     Reference: _____

7.   Why is GFCI protection for personnel using 120-volt, 15 and 20 amp rated receptacles, especially outdoors, not required for carnival and exhibition shows?

     Answer: _____

     _____

     _____

     _____

     Reference: _____

8.   I have been told that a tap conductor could not be tapped from a conductor that had been tapped from another conductor. Where is this stated in the *NEC*®?

     Answer: _____

     _____

     _____

     _____

     Reference: _____

9.   An individual 15 amp branch-circuit is installed for refrigeration equipment in the kitchen of a dwelling unit per Section 210-52(b)(1), Exception No. 2. Does *NEC*® Section 220-16(a) require additional 1,500 volt-amperes for this 15 amp circuit?

     Answer: _____

     _____

     _____

     _____

     Reference: _____

10. Do all the receptacle outlets in a hotel or motel room located conveniently for permanent furniture layout have to be accessible?

    Answer: _____

    _____

    _____

    _____

    Reference:_____

11. If outlets end up behind nonstationary furniture such as a recliner chair, bed table, or desk such that they would have to be moved in order to plug into a receptacle outlet, does this mean they are not considered accessible?

    Answer: _____

    _____

    _____

    _____

    Reference:_____

12. When should a transfer-switch switch the grounded conductor as well as the ungrounded conductors between a utility service and a generator service?

    Answer: _____

    _____

    _____

    _____

    Reference:_____

13. Can a receptacle be located below the rim of a pedestal sink in a bathroom and if so, how far below before it would not be considered as the required receptacle outlet for the bathroom basin?

    Answer: _____

    _____

    _____

    _____

    Reference:_____

14. The requirement for a single 20 amp circuit to supply the required receptacles in dwelling bathrooms in the 1996 *NEC®* does not permit any other loads to be connected to this circuit. Have these requirements changed for the 1999 *NEC®*?

    Answer: _____

    _____

    _____

    _____

    Reference: _____

15. Many large assembly and exhibition halls utilize cable trays for distribution of temporary wiring for an event. Is it permissible to use SO or SJT flexible cords in these cable trays for such temporary installations?

    Answer: _____

    _____

    _____

    _____

    Reference: _____

16. The Code requires that electric motors used on fire pumps be _____ for fire pump service.

    Answer: _____

    _____

    _____

    _____

    Reference: _____

17. Do the equipotential plane requirements of agriculture buildings apply to chicken houses?

    Answer: _____

    _____

    _____

    _____

    Reference: _____

18. If an equipment grounding conductor is not run with a 4-wire, 3-phase feeder to a second building, can the grounded conductor of the feeder circuit be grounded to the same contiguous metal water pipe as the main electrical service in the first building where that water line runs to the second building?

    Answer: _____

    _____

    _____

    _____

    Reference: _____

19. Many new dwellings have a conduit or tubing sleeve installed from the top of a panelboard up into the attic space so that future circuits can be gained from the panelboard without having to fish cables or cut sheetrock. Are such installations proper because future cables cannot be secured to the panelboard?

    Answer: _____

    _____

    _____

    _____

    Reference: _____

20. Are all 15- and 20-ampere receptacles for dwelling unit unfinished accessory buildings required to be GFCI protected?

    Answer: _____

    _____

    _____

    _____

    Reference: _____

1999 CODE CHANGE QUIZ THREE

1. Where multiple-conductor power cables are installed in parallel, is it necessary for the equipment grounding conductor in each cable to be sized per the feeder or branch-circuit overcurrent device protecting the circuit according to Table 250-122?

    Answer: _____

    _____

    _____

    Reference: _____

2. What is a "Power Panelboard?"

   Answer: _____

   _____

   _____

   _____

   Reference: _____

3. Does the Code allow splicing of service-entrance conductors?

   Answer: _____

   _____

   _____

   _____

   Reference: _____

4. Is it permissible to use NM cable with temporary wiring on a construction site?

   Answer: _____

   _____

   _____

   _____

   Reference: _____

5. Is it permissible to prewire electrical nonmetallic tubing before installing it in a concrete slab?

   Answer: _____

   _____

   _____

   _____

   Reference: _____

6. Is there a limit on how many individual branch-circuits are permitted to be run from one building to another on the same property under single management?

   Answer: _____

   _____

   _____

   _____

   Reference: _____

7. Why is MC cable allowed in assembly-type occupancies with over 100 occupants while Type AC cable is not allowed?

   Answer: _____

   _____

   _____

   _____

   Reference: _____

8. Do the requirements that state "*. . . receptacle outlets shall not be installed in a face-up position in the work surfaces or counter tops*" apply to wet bar sink locations?

   Answer: _____

   _____

   _____

   _____

   Reference: _____

9. What is the definition of a "Supervised Industrial Installation" as opposed to a commercial establishment?

   Answer: _____

   _____

   _____

   _____

   Reference: _____

10. All 15- and 20-ampere, 125-volt receptacles in pediatric wards, rooms, or areas must be _____ or employ a _____ cover.

    Answer: _____

    _____

    _____

    _____

    Reference: _____

11. Is it ever permissible to mount a small and lightweight lighting fixture to a device box, especially if the device box is a flush-mounted-type existing in a wall?

    Answer: _____

    _____

    _____

    _____

    Reference: _____

12. When trimming-in a job, how many inches of individual conductor is required to be left from the front of a 2 × 4 inch outlet box for the connection of or future replacement of toggle switches and receptacles?

Answer: _____

_____

_____

_____

Reference: _____

13. Does the *NEC*® permit NM cable to be used for lease space wiring in a four-story building with 6,000 square feet per floor that houses novelty shops?

Answer: _____

_____

_____

_____

Reference: _____

14. Where in the Code is a "curb" around the floor opening required for busways passing vertically pass through more than one floor of all buildings?

Answer: _____

_____

_____

_____

Reference: _____

15. How many ground rods are required regardless of the resistance when the rod is to supplement a metallic water line electrode?

Answer: _____

_____

_____

_____

Reference: _____

16. Does the new Article 830, "Network-Powered Broadband Communication Systems" cover high-power communication systems?

Answer: _____

_____

_____

_____

Reference: _____

17. Is at least one general care patient bed location receptacle outlet always required to be served by the normal system power supply, or can all general care patient bed receptacles be served through emergency system connected panelboards?

Answer: _____

_____

_____

_____

Reference: _____

18. How many "main power feeders" may a single-family dwelling have and do the loads they serve make any difference? For example, a large single-family dwelling has a 600 amp rated service with two 100 amp fused service switches, one 200 amp fused switch, and a 400 amp fused switch each feeding interior located panelboards. Are all "main power feeders?"

Answer: _____

_____

_____

_____

Reference: _____

19. Is GFCI protection for 120-volt receptacle outlets required on the third floor outdoor balcony of a dwelling building?

Answer: _____

_____

_____

_____

Reference: _____

20. Where in the Code are "manholes" covered?

Answer: _____

_____

_____

_____

Reference: _____

Examples of Maximum Length Equipment Grounding Conductor (Steel EMT, IMC, GRC, and Copper or Aluminum Wire) Computed as a Safe Return Fault Path to Overcurrent Device Based on 1997 Georgia Tech Software Version (GEMI 2.4, 1998) With an Arc Voltage of 40 and 4 IP at 25°C Ambient Based on a Circuit Voltage of 277 Volts to Ground THHN/THWN Insulation

Overcurrent Device Rating Amperes (75°C)	400% (4IP) Overcurrent Device Rating Amperes	Circuit Conductor Size AWG-kcmil Copper or Aluminum	Steel EMT, IMC, GRC Trade Size	(1) Equipment Grounding Conductor Size Copper or Aluminum	Length of EMT Run Computed Maximum (In Feet)	Length of IMC Run Computed Maximum (In Feet)	Length of GRC Run Computed Maximum (In Feet)	Copper Grounding Conductor w/o Steel Conduit Maximum Run (In Feet)	Aluminum Grounding Conductor w/o Steel Conduit Maximum Run (In Feet)
20	80	12	1/2	—	1170	1179	1140	—	—
20	80	12	—	12	—	—	—	890	—
20	80	10 AL	—	10 AL	—	—	—	—	870
30	120	10	1/2	—	—	1135 (4)	—	—	—
30	120	10	3/4	—	1199	1182	1143	—	—
30	120	10	—	10	—	—	—	946	—
30	120	8 AL	—	8 AL	—	—	—	—	920
40	160	8	3/4	—	1208	1228	1170	—	—
40	160	8	1	—	1326	1276	1239	—	—
40	160	8	—	10	—	—	—	871	—
40	160	8 AL	—	8 AL	—	—	—	—	690
60	240	6	3/4	—	1039	1134	1075	—	—
60	240	6	1	—	1197	1186	1131	—	—
60	240	6	—	10	—	—	—	676	—
60	240	4 AL	—	8 AL	—	—	—	—	657
100	400	3	1 1/4	—	1192	1176 (4)	1107	—	—
100	400	3	—	8	—	—	—	680	—
100	400	1 AL	—	6 AL	—	—	—	—	659
200	800	3/0	2	—	1157	1155	1077	—	—
200	800	3/0	—	6	—	—	—	598	—
200	800	250 AL	—	4 AL	—	—	—	—	578

Note: Software is not limited to above examples

(1) Per 1999 NEC® Table 251-18

Examples of Maximum Length Equipment Grounding Conductor (Steel EMT, IMC, GRC, and Copper or Aluminum Wire) Computed as a Safe Return Fault Path to Overcurrent Device Based on 1997 Georgia Tech Software Version (GEMI 2.4, 1998) With an Arc Voltage of 40 and 4 IP at 25°C Ambient Based on a Circuit Voltage of 120 Volts to Ground THHN/THWN Insulation

Overcurrent Device Rating Amperes (75°C)	400% (4IP) Overcurrent Device Rating Amperes	Circuit Conductor Size AWG-kcmil Copper or Aluminum	Steel EMT, IMC, GRC Trade Size	(1) Equipment Grounding Conductor Size Copper or Aluminum	Length of EMT Run Computed Maximum (In Feet)	Length of IMC Run Computed Maximum (In Feet)	Length of GRC Run Computed Maximum (In Feet)	Copper Grounding Conductor w/o Steel Conduit Maximum Run (In Feet)	Aluminum Grounding Conductor w/o Steel Conduit Maximum Run (In Feet)
20	80	12	1/2	—	395	398	384	—	—
20	80	12	—	12	—	—	—	300	—
20	80	10 AL	—	10 AL	—	—	—	—	293
30	120	10	1/2	—	358	383	364	—	—
30	120	10	3/4	—	404	399	386	—	—
30	120	10	—	10	—	—	—	319	—
30	120	8 AL	—	8 AL	—	—	—	—	310
40	160	8	3/4	—	407	414	395	—	—
40	160	8	1	—	447	431	418	—	—
40	160	8	—	10	—	—	—	294	—
40	160	8 AL	—	8 AL	—	—	—	—	232
60	240	6	3/4	—	350	383	363	—	—
60	240	6	1	—	404	400	382	—	—
60	240	6	—	10	—	—	—	228	—
60	240	4 AL	—	8 AL	—	—	—	—	221
100	400	3	1 1/4	—	402	397 (4)	373	—	—
100	400	3	—	8	—	—	—	229	—
100	400	1 AL	—	6 AL	—	—	—	—	222
200	800	3/0	2	—	390	389	363	—	—
200	800	3/0	—	6	—	—	—	201	—
200	800	250 AL	—	4 AL	—	—	—	—	195

(1) Per 1999 NEC® Table 251-18

Note: Software is not limited to above examples

Appendix 4: Answer Key

SAMPLE EXAM QUESTION REVIEW

1. A. Ref.—Article 351
2. B. Ref.—Section 347-14
3. C. Ref.—Section 349-10(b)
4. C. Ref.—Section 370-28(a)(2)
5. B. Ref.—Section 300(6)(c)
6. B. Ref.—Section 364-5
7. C. Ref.—Chapter 9, Table 1
8. B. Ref.—Sections 400-13 and 240-4, Exception 1
9. D. Ref.—Section 310-14
10. D. Ref.—Section 300-4(f)
11. D. Ref.—Section 336-18
12. A. Ref.—Table 310-15(b)(6)
13. D. Ref.—Section 300-5(d)
14. C. Ref.—Table 310-15(b)(6)
15. D. Ref.—Section 280-25
16. C. Ref.—Section 384-15
17. B. Ref.—Section 230-54(c) Exception
18. A. Ref.—Section 230-31(b)
19. B. Ref.—Section 230-9 Exception
20. C. Ref.—Section 230-24(a)
21. D. Ref.—Section 240-83(c)
22. D. Ref.—Section 110-26(d)
23. B. Ref.—Section 230-95
24. D. Ref.—General knowledge
25. B. Ref.—Table 430-37
26. A. Ref.—Section 430-81(c)
27. B. Ref.—Section 380-8(b)
28. C. Ref.—Section 450-43(a)
29. A. Ref.—Section 300-7(a)
30. B. Ref.—Section 410-66(a)
31. C. Ref.—Section 410-76(c)
32. A. Ref.—Section 700-12(b)(2)
33. C. Ref.—Section 520-25(a) and (c)
34. D. Ref.—Section 501-5(a)
35. A. Ref.—Section 500-8
36. D. Ref.—Section 511-6(a)
37. B. Ref.—Section 517-18(b)
38. C. Ref.—Sections 675-15, 250-106 (FPN) #2
39. A. Ref.—Sections 440-61, 440-3, and 250-114(3)(a)
40. B. Ref.—Section 250-66(a)
41. D. Ref.—Section 250-50(a)
42. C.
43. A. Ref.—Sections 210-52, 210-11(c)
44. B. Ref.—Section 230-95
45. C. Ref.—Chapter 9, Tables 1, 4, 5
46. C. Ref.—Section 424-3(b)
47. B. Ref.—Chapter 3
48. D. Ref.—Section 90-2(b)
49. C.
50. C. Ref.—Section 460-2

QUESTION REVIEW CHAPTER 2

1. No. However, there is an exception: Section 351-23(b) for electrical signs. Ref.—Section 351-23(b) Exception

2. Governmental bodies exercising legal jurisdiction over the electrical installation, insurance inspectors, or others with the authority and responsibility for governing the electrical installation. Ref.—Section 90-4

3. Ships, railway rolling stock, aircraft, or automotive vehicles. Ref.—Section 90-2(b). Four installations are not covered, for example, see Section 90-2(b)(1)-(5).

4. Fine print notes (FPN). Ref.—Section 90-5(c)

5. Section 110-3(b)

6. When adopted by the regulatory authority over the intended use. Ref.—Pages 70-i, which states the Code is advisory as far as NFDA and ANSI are concerned but is offered for use in law for regulatory purposes.

7. No. It is the intent that all premise wiring or other wiring of the utility-owned meter read equipment on the load side of service point of buildings, structures, or other premises not owned or leased by the utility are covered. It also is the intent that buildings used by the utility for purposes other than listed in Section 90-2(b)(5), such as office buildings, warehouses, machine shops, or recreational buildings, that are not an integral part of the generating plant, substation, or control center are also covered. Ref.—Section 90-2(b)(5)

8. There has been a change in this edition of the *NEC*® from the wording of the previous edition.

9. (a) 1999, (b) September, 1998.

10. 1881, The National Fire Engineers met in Richmond, Virginia, which resulted in the first document that led to the *National Electrical Code*®. Various meetings were held after that; then in 1897 some various allied interests of the church, electrical, and architectural concerns met, and in 1911 NFPA was set up as the sponsor.

11. Yes. Ref.—Article 305, Sections 305-8, 305-4(c)

12. Chapter 8. Ref.—Section 90-3

13. No, it is not a training manual for untrained persons. Ref.—Section 90-1(c)

14. No, only underground mines and rolling stock for surface mines. All surface mining other than the rolling stock are covered by the *NEC*®. Ref.—Section 90-2(b)(2)

15. Nowhere in the Code does it state that it is the minimum allowed. Only when adopted by a local jurisdiction would it be considered the minimum of that jurisdiction. When an installation is installed for the *National Electrical Code*®, it provides a safeguarding of persons and property, and the provisions are considered necessary for safety and compliance with the *NEC*® will result in an installation essentially free from hazards. However, it may not be efficient or convenient or adequate in some installations. Ref.—Section 90-1(a) and (b)

16. Definitions, Article 100, to continue for three hours or more

17. 1993. The *National Electrical Code*® is adopted every three years; there being a 1993, a 1996, and a 1999. Therefore, not adopting the two editions, they would be using the 1993 *NEC*®.

18. Article 725 covers Class 2 wiring.

19. Article 517, Health Care Facilities, Section 517-3, Definitions

20. Section 250-50(a)(2). The water pipe must be used if available, and supplemented with another grounding electrode of one of the types listed in Section 250-50 or 52.

QUESTION REVIEW CHAPTER 3

(Reduce fractions to their lowest terms.)

1. ½

2. ¼

3. 1/10

4. ⅖

5. ¾

6. ¾

7. ½

8. ⅗

9. ⅗

10. ⅕

11. ¼

12. ¼

(Change mixed numbers to improper fractions.)

1. 3/2

2. 13/4

3. 43/8

4. 17/4

5. 57/8

6. $^{13}/_6$

7. $^{23}/_4$

8. $^{25}/_3$

9. $^{46}/_7$

10. $^{29}/_8$

(Change improper fractions to mixed numbers.)

1. $4\frac{1}{2}$

2. $2\frac{2}{5}$

3. $12\frac{4}{5}$

4. $3\frac{1}{4}$

5. $9\frac{2}{3}$

6. $15\frac{1}{2}$

7. $1\frac{2}{3}$

8. $4\frac{1}{3}$

(Multiply whole numbers and fractions. Answer in proper reduced form.)

1. 4

2. $^3/_8$

3. $^1/_{40}$

4. $^9/_2 = 4\frac{1}{2}$

5. $^{21}/_4 = 5\frac{1}{2}$

6. $^5/_2 = 2\frac{1}{2}$

7. 9

8. $^{15}/_7 = 2\frac{1}{7}$

(Convert fractions to decimal equivalents.)

1. .25

2. .625

3. .75

4. .866

5. .375

(Resolve to whole numbers.)

1. 16

2. 81

3. 2,500

4. 1

5. 1,000

(Find the square root.)

1. 5

2. 9

3. 7

4. 56

5. 101

6. 29

7. 78

(Solve the following problems.)

1. .05.   Ref.—Ohm's Law

2. 100%.   Ref.—Chapter 3

3. 5 amperes.   Ref.—Chapter 3, Ohm's Law

4. 1.33 ohms.   Ref.—Chapter 3

5. 14.4 ohms.   Ref.—Chapter 3

6. 6 volts.   Ref.—Chapter 3

7. 33.34 ohms

8. 83.34 ohms

9. 23.81 ohms

10. 107.15 ohms

11. 37.24 volts

12. 56 volts

13. 26.66 volts

14. 1.12 amps

QUESTION REVIEW CHAPTER 4

1. Yes.   Ref.—Section 90-2a

2. Chapter 8. Article 250 does not apply in Chapter 8 unless specifically referenced.

3. Article 410, Part L

4. False. Branch-circuits are referenced in many articles in the *NEC*®.

5. Article 516

6. Article 422 for appliances, specifically Section 422-2. Electrical heating is covered in Article 424 and heat pumps in Article 440.

7. Chapter 9, Tables 1 through 8 and all Notes

8. Lighting and power in mine shafts are not covered by the *NEC*®, Article 90, Section 90-2B

9. Section 300-50, Article 300

10. Article 225, which covers outside branch-circuits and feeders, Section 225-8

11. Chapter 9, Table 10

12. Places of assembly, Article 518

13. *NEC*® Article 250, Section 250-34

14. Article 680, Section 680-22, Part B

15. Article 90, Section 90-3

16. These requirements appear in the building codes or the NFPA Life Safety Code; they are not found in the *NEC*®.

17. Article 760

18. Yes. Ref.—Section 305-6(a)

19. Article 502—(a) Class 2, (b) Division 1. Article 500 for the definitions and Article 502 for the specific requirements.

20. It is a violation to put over 1,000-volt wiring in a residence. Neon signs requirements are found in Article 600. However, Article 410 requirements limit installation under 1,000 volts in dwellings. Ref.—Section 410-80(b).

QUESTION REVIEW CHAPTER 5

1. 4 feet. Ref.—Section 110-26, Table 110-26(a)

2. Two. Ref.—Section 110-26(c)

3. Yes. Ref.—Section 110-26(d)

4. 6½ feet. Ref.—Section 110-26(e)

5. One. Where the required working space is doubled, only one door is required. Ref.—Section 110-26(c), Exception 2.

6. No. The required disconnecting means and supplementary overcurrent device must have a working clearance in accordance with Section 110-26. Clear working clearance of 30 inches wide and 40 inches deep are required from the ground level up 6½ feet.

7. 30 inches. Ref.—Section 110-26(a)

8. Ungrounded. Ref.—Section 240-20

9. 60%. Ref.—Chapter 9 Table Notes

10. Clothes closet. Ref.—Section 240-24d

11. 1¼ inches. Ref.—Section 300-4(a)(1)

12. 600 volts. Ref.—Article 331-4(6)

13. Yes. Ref.—Section 230-43(16)

14. Yes, it shall be bonded to the electrode at each end and all intervening raceway boxes and enclosures between the service equipment and the grounding-electrode conductor. Ref.—Section 250-92(a)(3)

15. Overcurrent device. Ref.—Section 250-122

16. Two. Ref.—Section 210-52(e)

17. Yes. Ref.—Section 680-73, except manufacturer's instructions may require Section 110-3(b)

18. Yes. Ref.—Section 210-52(a)

19. No. Ref.—Section 210-52(a) for site-built dwellings; Section 550-8(d) Exception 4 for mobile homes

20. Section 210-63. Ref.—Article 210

QUESTION REVIEW CHAPTER 6

1. 10, 80% loading. Ref.—Section 220-10(b)

2. 1,120 square feet. Ref.—Chapter 3

3. 2. One additional 20 amp circuit must be added for bathroom receptacles. Ref.—Table 220-3(a) and Footnotes, also see Figure 6–4, Section 220-11(c)(3)

4. 5—2 kitchen appliance circuits, 1 laundry circuit, 1 bathroom receptacle circuit, and 1 heating circuit Ref.—Sections 422-12, 210-11(c)

5. 2. Ref.—Section 210-11(c)(1)

6. Yes. Ref.—Section 210-11(c)(2)

7. 30 amperes. Section 422-10

8. None. The *NEC®* does not require power to the structure.   Ref.—Section 210-52(g)

9. 2.   Ref.—Section 210-52(e)

10. Four minimum, one for the required bathroom circuit, one for required outdoor receptacles and two for the required kitchen appliance circuit minimum. Ref.—Section 210-8(a), also see Section 210-52 and all subsections.

11. Yes.   Ref.—Section 210-70(a)

12. No.   Ref.—Section 210-63 and 70

13. Yes.   Ref.—Section 210-63

14. 15 or 20 amperes.   Ref.—Section 210-19(d)

15. No.   Ref.—Section 422-12, Definition of Individual Circuit

16. No.   Ref.—Section 410-8

17. Yes.   Ref.—Section 210-52(b)(2) Exception #2

18. Yes.   Ref.—Definition of direct grade access, Section 210-8

19. No, 2 receptacles at direct grade access are required.   Ref.—Sections 210-52 and 210-8

20. Section 230-79(c)

    (a) Section 220-3(a) and Table 220-3(a); 28 × 40 = 1,120 sq. ft. 3 VA × 1,120 = 3,360 volt-amperes

    (b) Section 220-16(a); 3,000 volt-amperes

    (c) Section 220-3(b); 4,500 volt-amperes

    (d) Section 220-16(b); 1,500 volt-amperes

    (e) Section 220-18; 5,000 volt-amperes

    (f) 0. Electricity not required; however, if supplied, the load would have to be included in calculation

    (g) Section 220-3(b); size not stated, therefore, a 20-ampere branch-circuit is recommended

    (h) 0. Section 220-3(b); Section 210-52(b) Example 2 could be general lighting or on appliance load

    (i) 0. None

    (j) 0. Gas range igniter can be fed from appliance circuit, Section 210-52(b), Exception #2

    (k) 0. None, load included in general lighting calculation, Section 220-3.

QUESTION REVIEW CHAPTER 7

1. 6.   Ref.—Section 230-208

2. A disconnecting means is required at each building.   Ref.—Article 225, Part B

3. Yes, you are limited to not more than six disconnects on this secondary side.   Ref.—Section 450-3—conductors must be protected in accordance with Section 240-3.

4. No, however, Section 250-54(e) requires the conductors be separated where of a different potential.

5. No. Section 240-24(e) prohibits branch-circuit overcurrent devices from being located within a residential bathroom.   Ref.—Section 240-24(e)

6. Yes. SE cable can be used in the same manner as Type NM cable, however, it is not considered NM cable and must comply with Articles 300 and 336.   Ref.—Section 338-4(a)

7. Usually no.   Ref.—Section 230-9. Yes, if the window is not designed to be opened, or where the conductors run above the top level of the window. Where windows are designed for opening and closing, then a minimum of 3 feet from the windows, doors, porches, fire escapes, or similar locations is required.

8. (a) Yes.   Ref.—Section 250-64(c)

    (b) They shall be sized in accordance with Section 250-66 for the largest conductor entering each respective enclosure.

9. Requirement to open all of the conductors is a disconnect requirement and not an overcurrent requirement.   Ref.—Section 240-20(b) gives the requirements for circuit breakers, Section 220-40 gives the requirements for fuses. Section 210-4 is the requirement for multiwire branch-circuits. Disconnecting, not overcurrent, is the purpose.

10. A separately derived system is defined in Article 100, and the grounding installation requirements are found in Section 250-30.   Ref.—Article 100, Section 250-30

11. Section 450-3—not more than six disconnects are permitted on the secondary side of the transformer.

12. No. Identification of the hi-leg is only required where the neutral is present. Ref.—Section 215-8, 230-56 and Section 384-3(e)

13. No. The grounding-electrode conductor must be continuous and the connection shall be made solidly and used for no other purpose. Ref.—Section 250-60

14. Less than ¼ inch. Ref.—Section 347-9 Exception

15. Grounded effectively is defined in Article 100 and the authority having jurisdiction will make that determination.

16. Article 336, nonmetallic sheath cable. Although Romex is a trade name, the term can be found in the index of the *NEC®*.

17. Article 349 covers the requirements of flexible nonmetallic tubing. It is manufactured in sizes ½ inch and ¾ inch only. *Note:* This is a flexible metallic tubing that is liquidtight without a non-metallic jacket; rarely used in construction installations.

18. Article 380

19. Chapter 9, Table 1, Footnote 9 states that multi-conductor cable of two or more conductors shall be treated as a single conductor cable, and cables that have an elliptical cross-section, the cross-section layering calculation shall be based on using the major diameter for the ellipse as the circle diameter.

20. Article 511 does not apply. Sections 511-1 and 511-2 clearly exempt these parking structures. Therefore, Chapters 1 through 4 apply, and the wiring method would be taken from those four chapters.

QUESTION REVIEW CHAPTER 8

1. Yes. Ref.—Section 240-20(b). However, handle ties are required.

2. No maximum length needed, provided they are outdoors. Ref.—Section 240-21(5)

3. No. Ref.—Section 240-60(a). Voltage to ground can exceed 300 volts on a wye connection and, therefore, would not be permitted.

4. Is not a limiting factor. However, six disconnect rule applies. Ref.—Section 230-90(a) Exception 3.

5. 100%. Section 220-15

6. No. Ref.—Section 501-4, 351-9 Exception 2, 501-16(b), and must comply with Section 250-102

7. Yes. Both the overcurrent device and the conductors would have to meet these requirements. These are Exceptions to the general rule. Ref.—Section 215-23

8. There is no limit. Ref.—Table 220-3(b)(10) *Note:* Many local jurisdictions set limits on 15- and 20-ampere branch-circuits in all installations. However, the *NEC®* does not specify a limit on dwelling branch-circuits for general use electrical outlets.

9. No. Ref.—Section 220-3(b)(9), 220-3(c)(6), also the obelisk at the bottom of 310-16 limits No. 12 conductors to a 20-ampere branch-circuit. 16×180 VA $= 2,880 \div 120$ V $= 24$ amperes.

10. (a) No. Ref.—Section 220-21

    (b) Yes. Ref.—Section 424-3(b)

11. No, only if installed in cable with 12 inches of cover. Otherwise, can be installed at the greater depth or in raceways. Ref.—Section 300-5 and Table 300-5

12. Yes. Ref.—Section 240-20(b)

13. No. Ref.—Section 215-2. Footnote: Branch-circuits are required to be calculated at 125% for water heaters—Section 422-13

14. Yes. Ref.—Section 210-19

15. No. Ref.—Section 210-6—12 amperes or less must be 120 volts; over that can be 240 volts. 1,400 VA $\div$ 240 V $= 5.833$ amperes.

16. 180 VA. Ref.—Sections 220-4, 220-3(b)

17. Yes, it complies with the tap rules in Section 240-21(b) for the 10-foot tap rules.

18. Yes, it complies with the 25-foot tap rule. Ref.—Section 240-21(b).

19. No, it does not comply with any of the tap rules in Section 240-21.

20. True. Ref.—Section 364-11

QUESTION REVIEW CHAPTER 9

1. Soldered.   Ref.—Section 250-8

2. 14.   Ref.—Section 600-7

3. Not required to be accessible.   Ref.—Section 250-68 Exception

4. White or natural gray.   Ref.—Section 200-6

5. 4/0.   Ref.—Section 250-24

   3-500 kcmil = 1,500 kcmil or 3 each 500 kcmil = 1,500 kcmil × 12½% = 187.5 kcmil

   1,500 × 12.5% = 187.5 kcmil

   Chapter 9, Table 8—3/0 167.8 kcmil too small; therefore, use 4/0 211.6 kcmil

6. 4.   Ref.—Table 250-22

7. Yes, bonding is not required.   Ref.—Section 250-94(3)

8. They are required to be run in each nonmetallic raceway and are required to be sized in accordance with the sum of the paralleled phase conductors.   Ref.—Sections 215-6 and 250-122 and Table 250-122.

9. Yes.   Ref.—Section 250-112(l)

10. No, the language is specific.   Ref.—Section 250-30

11. Yes, it must be bonded. No, it is not sufficient to bond at one end. It must be bonded at both ends and at all enclosures in which it passes through. Ref.—Section 250-92(a)(3)

12. No, a driven rod is not acceptable. An equipment grounding conductor must be carried with the circuit.   Ref.—Section 551-76(a)(3)

13. No, the requirements for separately derived transformers are found in Section 250-30. These requirements do not amend the requirements of Section 250-50, but they are the specific requirements. Therefore, no supplementary grounding electrode is required.

14. No. 8 AWG copper or equivalent.   Ref.—Section 551-56(c)

15. Yes.   Ref.—Section 250-104. *Note:* The interior gas piping system is required to be grounded for safety. It can be grounded with the circuit conductor equipment ground most likely in which it would come in contact. For example, a central heating electric furnace may have No. 12 conductors supplying the fan motor. The No. 12 equipment grounding conductor run with those circuit conductors could be the conductor used to bond the gas piping system in this case. **Warning:** The underground gas piping system may not be grounded. Electrolysis may occur, creating dangerous underground gas leaks.

16. Yes, a No. 1 copper or a No. 2/0 aluminum or copper-clad aluminum wire is required in each conduit.   Ref.—Table 250-122 and Section 250-122, which states that where run in parallel in multiple raceways or cables as permitted, the equipment grounding conductors also shall be run in parallel, each equipment grounding conductor shall be sized on the basis of the ampere rating of the overcurrent device protecting the circuit conductor in the raceway in accordance with Table 250-122.

17. No.   Ref.—Section 110-14(b)

18. No. 8.   Ref.—Table 250-122. This is a minimum size conductor and may have to be larger if the fault current available exceeds that for which the No. 8 conductor will carry.

19. Yes. The flexible metal conduit cannot exceed 6 feet for the combined two pieces of flexible metal conduit.   Ref.—Section 250-62

20. The maximum rating of the grounding impedance, no smaller than No. 8 copper or No. 6 aluminum.   Ref.—Section 250-24, 250-36

QUESTION REVIEW CHAPTER 10

1. Within 12 inches of the outlet or box, and every 4½ feet.   Ref.—Section 338-4 and Section 336-15. The 1999 *NEC*® requires a separate equipment grounding conductor for all new installations. Therefore, Type SE cable may no longer be appropriate, Type SER would be acceptable.

2. Maximum number of conductors shall be permitted using the provisions of Section 370-16(b). Conduit bodies shall be supported in a rigid and secure manner.   Ref.—Section 370-16(c)

3. Yes. Ref.—Section 351-24

4. Yes. Ref.—Section 345-12 Exception

5. Yes. Ref.—Section 331-11 Exception—Metal studs are the required support when they do not exceed 3 feet apart.

6. NM cable shall be supported at intervals not exceeding 4½ feet and within 12 inches of every cabinet, box, or fitting. Two conductor cables shall not be stapled on edge, run through holes in wood studs or rafters shall be considered support. Ref.—Section 336-18

7. No. Ref.—Section 362-16

8. Aboveground conductors must be installed in a rigid metal conduit, intermediate metal conduit, rigid nonmetallic conduit in cable tray or busway or cable bus or in other identified raceways, or as open runs of metal clad cable suitable for the use. Open runs of Type MB cable, bare conductors or bare bus bars also are permitted as they apply. Ref.—Section 300-50

9. Yes, provided that carpet squares not exceeding 3×3 are used in a wiring system in accordance with Article 328 Type FCC cable and associated accessories are used. Ref.—Article 328

10. Metallic wireway shall be established in accordance with Section 362-5 and shall not exceed 30 current-carrying conductors at any cross-sectional area. The sum of cross-sectional areas of all conductors at any cross-section of the wireway shall not exceed 20%. The derating factors in Table 310-15(b)(2) do not apply. For nonmetallic wireway, Section 362-19 applies. The cross-sectional area contain conductors at any cross-section of the nonmetallic wireway shall not exceed 20% of the interior cross-sectional area. The derating factors in Table 310-15(b)(2) do apply to the current-carrying conductors at the 20% specified. *Note:* This is a critical difference. Caution should be used when reading questions as they apply to a specific wiring method, metallic or nonmetallic, because the conditions and rules are very different.

11. 1¼ inches from the nearest edge of the stud. Ref.—Section 300-4(d)

12. There is no difference in the permitted uses or installation requirements in the two products. Article 345 and Article 346 are essentially the same. Ref.—Article 345 and Article 346

13. 1/0. Ref.—Section 300-4

14. Yes, in any building exceeding three floors, nonmetallic tubing must be concealed within a wall, floor, or ceiling where the material has at least a 15-minute finished rating as identified in the listings of the fire-rated assemblies. Ref.—Section 331-3(2)

15. They must be removed. They are not permitted for a period to exceed 90 days. Ref.—Section 305-3(b)

16. Lengths not exceeding 4 feet can be used to connect physically adjustable equipment and devices that are permitted in the duct or plenum chamber. Ref.—Section 300-22(b)

17. ¼ inch. Ref.—Section 347-9 Exception

18. Yes, even though Section 300-3 requires that all conductors of the same circuit, including the neutral and equipment grounding conductors, must be run within the same raceway, cable trench, cable, or cord; Exception 2 to that section permits that with column type panelboards; the auxiliary gutter, and pull box may contain the neutral terminations. Ref.—Section 300-3(b)(4)

19. No, each cable shall be secured to the panelboard individually. Ref.—Section 373-5(c)

20. No, a branch-circuit or a lighting panelboard is one that 10% of its overcurrent devices are rated 30 amperes or less for which neutral connections are provided. A lighting and branch-circuit panelboard, as required by the Code, shall have not more than two circuit breakers or two sets of fuses for disconnecting. Ref.—Section 384-14 and Section 384-16—A disconnect is required ahead of the lighting and branch-circuit panelboard although the load is relatively small in this cabin.

QUESTION REVIEW CHAPTER 11

1. Section 430-6a. Ref.—Section 430-6a

2. 100 ampere. Ref.—Section 430-6, Tables 430-150 and 430-152

3. 175%. Ref.—Tables 430-150 and 430-152

4. No. 14.   Ref.—Table 210-24

5. 20.   Ref.—Sections 240-3, 240-6 and Table 310-16

6. 40.   Ref.—Sections 240-3, 240-6 and Table 310-16

7. 100.   Ref.—Sections 240-3, 240-6, and Table 310-16

8. 200.   Ref.—Sections 240-3, 240-6 and Table 310-16

9. 125.   Ref.—Section 450-3(b)

10. 175%.   Ref.—Table 430-152

11. 125%.   Ref.—Article 430

12. 6.   Ref.—Table 430-91

13. Indoors.   Ref.—Section 410-73(e) and (f)

14. Fire protection or equipment cooling.   Ref.—Section 450-47

15. Accessible.   Ref.—Section 450-13(a)

16. 3.   Ref.—Section 460-2

17. 8.   Ref.—Section 440-32 and Table 310-16

18. 112.5 kVA.   Ref.—Section 450-21(b)

19. 600 watts.   Ref.—Ohm's Law (*Clue:* Resistance does not change.)

20. 87%.   Ref.—Chapter 3

QUESTION REVIEW CHAPTER 12

1. Nonconductive, conductive, composite.   Ref.—Section 770-5

2. True.   Ref.—Section 511-6

3. Yes, where the raceway emerges from the ground, at both the office and at the dispenser.   Ref.—Section 514-6a and b

4. False.   Ref.—Section 700-1

5. Class 1, Division 2.   Ref.—Section 511-3(a)

6. Parallel interconnection.   Ref.—Section 700-6

7. ⅝.   Ref.—Section 501-5(c)(3)

8. 12 inches.   Ref.—Section 490-24

9. Two duplex or four single receptacles.   Ref.—Section 517-18(b)

10. 6.   Ref.—Section 220-10(a) and (b), Tables 430-148 and 300-16

11. 15 feet.   Ref.—Section 600-10(d)(2)

12. False

13. True

14. False.   Ref.—240-54(b)

15. True.   Ref.—Bussmann SPD Protection Handbook

16. (a) Electrician or electrical contractor

    (b) Field marking is required to state "Caution—Series Rated System, ___ Amperes Available. Identified Replacement Components Required."   Ref.—Section 110-22.

    (c) Manufacturer

17. 6 amperes.   Ref.—$I = \dfrac{HP \times 746}{E \times PF \times eff\%}$

    Note: When power factor is not noted it must be calculated at 100%.

18. 60.   Ref.—Tables 430-150 and 152

19. 27.   Ref.—Ohm's Law, Chapter 3

20. 18.   Ref.—Ohm's Law, Chapter 3

QUESTION REVIEW CHAPTER 13

1. Yes.   Grounding electrode.   Ref.—Sections 810-20(c), 810-21

2. No. Chapter 8 stands alone. There are no burial depth requirements for satellite dishes installed anywhere there is no safety hazard, and, therefore, the only consideration is for the customer's protection of his or her own equipment, *NEC®* Section 800-10. *Note:* Also look at Section 90-3. Chapter 8 covers communication systems and is independent of the other chapters, except where they are specifically referenced. Chapter 8 stands alone, and other chapters cannot be applied unless specifically referenced.

3. Yes.   Ref.—Section 800-33

4. 6 feet.   Ref.—Section 800-13

5. It is an intrinsically safe circuit covered in Article 504 of the *National Electrical Code®*.

6. Article 410-56c covers these isolated ground receptacles intended for the reduction of electrical noise. 1999 *NEC®* change requires receptacles used for "isolated ground" be marked on the face of the receptacle with an orange triangle. The solid orange receptacle is no longer acceptable.

7. Yes.   Ref.—Article 650

8. Yes, Article 110, Section 110-26 for installations of less than 600 volts, Section 110-32 for installations over 600 volts.

9. Yes. Section 100-12c clearly states that paint, plaster, and other abrasives may damage the bus bars, wiring terminals, insulators, and other surfaces in panelboards. Therefore, the internal parts of this equipment must be covered and protected.

10. Inspector was correct. Section 550-23 requires the service equipment to be located inside and not more than 30 feet from the exterior wall of the mobile home. An exception permits the installation of the service equipment on the mobile home when installed by the manufacturer. Electrical contractors, however, cannot make this installation.

11. Optional standby system is covered in Article 702 and is defined in 702-2.

12. Color coding is not specifically required; however, Section 210-4d requires the identification of ungrounded conductors on multiwire branch-circuits under certain conditions. This identification can be by color coding, marking tape tagging, or other equally effective means.

13. Section 225-7b covers outdoor branch-circuits for lighting equipment and there is no maximum number. See Section 215-4(a) for feeders.

14. Yes. Article 547, Agriculture Buildings, covers this type of installation. See also Section 250-32(e).

15. Yes. Article 328 covers this type installation. However, the carpet must be laid in squares not to exceed 3 × 3.

16. Article 280, Surge Arrestors, covers that type of installation.

17. Article 305 permits a class less than would be required for permanent installation. However, there are time constraints. *Note:* 305-4a requires the service on that temporary pole to comply with Article 230.

18. 86%. Ref.—Chapter 3, Duff & Herman, *Alternating Current Fundamentals, 4E* by Delmar, Thomson Learning

19. 60 Hp.   Ref.—Loper, *Direct Current Fundamentals, 4E* by Delmar, Thomson Learning

    $$EFF \frac{W}{\sqrt{3} \times E \times I}$$

20. 0.105.   Ref.—Loper, *Direct Current Fundamentals, 4E* by Delmar, Thomson Learning

PRACTICE EXAM 1

1. Yes.   Ref.—Section 511-10

2. No.   Ref.—Section 680-71, however, Section 410-4(a) and (d) limits the use of pendant-type fixtures over this area.

3. Yes.   Ref.—Section 210-7(d)(3) and no, Ref. Section 210-7(d), but they must be marked "GFCI" protected.

4. No.   Ref.—Section 210-8(a)(6)—only those serving the counter tops. *Note:* Only very old ranges have receptacles mounted on them.

5. No.   Ref.—Section 370-3, Exceptions 1 and 2

6. Yes.   Ref.—Section 410-16(h)

7. 15-minute finish rating.   Ref.—Section 331-3(3)

8. No.   Ref.—Section 300-4(b)(2)

9. Where run through the framing members ENT shall be considered supported.   Ref.—Section 331-11

10. Yes.   Ref.—Section 300-4b

11. Yes.   Ref.—Sections 339-3(a)(4) and 336-26

12. No.   Ref.—Section 517-10, Exceptions 1 through 3

13. Yes.   Ref.—Section 210-52a

14. No.   Ref.—Section 300-13b

15. Yes. Ref.—Sections 210-52(c) and 210-8(a)(6) *Note:* The GFCI requirement would not apply if the receptacle was located behind the refrigerator.

16. No. Ref.—Section 210-52(a) and (e)

17. No. The requirement is now 6½ feet. Ref.—Section 380-8(a)

18. See referenced exceptions and general rule. Ref.—Section 410-31

19. General and critical care patient bed location. Ref.—Section 517-18(b)

20. No. Ref.—Section 210-52(a)—shall be in addition to those required if over 5½ feet above the floor. See Section 210-8(a)(7)

21. Self-closing. Ref.—Section 410-57(b)

22. No. Ref.—Section 680-25(b)

23. Yes, provided it is installed and supported in accordance with Article 336. Ref.—Section 336-3, 4, 10, and 15

24. No, they may be rubber or thermoplastic types. Ref.—Section 225-4

25. Yes. Ref.—Section 410-31

PRACTICE EXAM 2

1. No. Ref.—Only as environmental conditions require—Sections 547-1 and 547-7a, b, c

2. No. Ref.—Sections 333-10, 300-4

3. Extra hard usage portable power cables listed for wet locations and sunlight resistance. Ref.—Section 553-7(a) and (b)

4. Yes. Ref.—Section 680-25(c)—NM cable has a covered or bare equipment ground conductor, not insulated, but is allowed within the dwelling occupancy.

5. No. Ref.—Section 680-21

6. No, water heaters are not listed with an attachment cord. Ref.—Section 400-7 and 8(1)

7. No, generally. Ref.—Section 373-8

8. Voltage with the highest locked rotor KVA per horsepower. Ref.—Section 430-7(b)(3)

9. Yes. Ref.—Section 680-25(d)

10. Yes. Ref.—Section and Tables 210-21(b)(2) and (b)(3)

11. Yes. Ref.—Section 210-23(a)

12. Not enclosed, buried, or in raceway. Ref.—Table 310-17

13. No. Ref.—Section 347-2 (FPN) is not mandatory, only explanatory

14. No, GFCI protection is still required. Ref.—Section 305-6(a)

15. Yes. Ref.—Section 210-8(a)(7)

16. No. Ref.—Section 210-8(a)(7)

17. No maximum. Ref.—Table 220-3(b)(10)

18. 6. Ref.—Section 517-18(b)

19. Yes. Ref.—Sections 430-81(c), 430-42(c)

20. Yes. Ref.—Section 230-70(a), Exception 1

21. Yes, one or more supplied by a 20-ampere current. Ref.—Sections 210-52(f)

22. No. Ref.—Section 410-17, 18, 20, and 21

23. Yes, flexible metal conduit must be bonded to grounded conductor. Ref.—Section 230-43(15)

24. No, only interior metal water piping systems. Ref.—Section 250-80(a)

25. No, generally. Ref.—Sections 370-22, 370-16(a)

PRACTICE EXAM 3

1. No. Ref.—Section 370-27(b)

2. No. Ref.—Section 370-16(a) and (b)

3. No. Ref.—Section 210-52(e) only single- and two-family dwellings

4. No, polarized or grounding type. Ref.—Section 410-42(a)

5. Yes. Ref.—Section 210-52(a)

   Note: Due to the arrangement of furniture, the receptacle located behind the door may be the only receptacle accessible for frequent use such as vacuuming.

6. No. Ref.—Section 305-4(g)

7. 6, where integrated clamps, etc., are not used. Ref.—Table 370-16(a)

8. Table 370-16(b)—370-16(a)(2)

9. Yes. Ref.—Section 510-4(a)

10. No, however, it is a raceway. Ref.—Article 348

11. Yes, up to 5 feet. Ref.—Section 348-12 Exception

12. Yes. Ref.—Section 410-16(c)

13. No. Ref.—Definitions, Article 100, and Section 410-4

14. No, must be spaced 1½ inches from the surface. Ref.—Section 410-76(b)

15. Yes, with exceptions. Ref.—Section 410-65(c)

16. No, only where the grounded (neutral) conductor is present. Ref.—Sections 215-8, 230-56, and 384-3(e)

17. Yes, all kitchen counter top receptacles are required to be GFCI protected. Ref.—Section 210-8(a)(6)

18. Generally no. Ref.—Section 373-8 and Exceptions

19. Yes. Ref.—Section 250-112(j), Part E Article 410

20. No. Ref.—Section 700-12(e)

21. No. Ref.—Section 210-8(a)(2) Exceptions 1 and 2

22. Yes "all." Ref.—Section 210-8(a)(1)

23. No; however, it cannot be counted as one of the counter top receptacles as required by Section 210-52(c). Ref.—Section 210-52(b) Exception 3

24. No. Ref.—Sections 370-1 and 410-64

25. Within 12 inches of service head, goose neck, and at intervals not exceeding 30 inches. Ref.—Sections 338-2, 230-51(a)

PRACTICE EXAM 4

1. 24 inches. Ref.—Table 300-5

2. Yes. Ref.—Table 210-21(b)(2)

3. No, you are not required to locate required receptacles behind furniture. However, the number required by Section 210-52(a) must be installed convenient to furniture arrangement. Ref.—Section 210-60 and Exception. 1999 *NEC®* requires all receptacles to be readily accessible.

4. Yes, but only if the box is listed for fan support. Ref.—Section 370-27(c) and Exception, Section 422-18

5. Ref.—Sections 410-31 and 410-32

6. Yes, generally. Ref.—Section 410-65(c) Exceptions 1 and 2

7. Yes. Ref.—Section 210-6(c)

8. Yes, it is required. Ref.—Section 700-12(e)

9. 150 kVA or less. Ref.—Diagram 517-41(3)

10. Yes. Ref.—Section 230-95

11. Yes, there is no exception for 15-volt lighting fixtures. Ref.—Section 680-41(b)(i) and Exceptions. 1999 *NEC®* see new Article 411.

12. Yes. Ref.—Section 348-1

13. No, it is not permitted. Ref.—Section 250-58

14. No. Ref.—Sections 110-26

15. Yes. Yes. Ref.—Section 210-8(a)

16. See Sections 547-8(a) and 250-32

17. No. Ref.—Part E, Article 230, specifically Section 230-70. *Note:* 1996 *NEC®* has added Section 225-8. If service is properly established at the pole, the answer to this question is Yes.

18. 24 inches. Ref.—Table 300-5

19. No, unless the cord and plug are part of listed appliance. Ref.—Sections 110-3(b), 400-8(1)

20. (a) No minimal burial depth requirements. Ref.—Article 810

    (b) As per Sections 810-21 and 810-15

21. Yes, provided they control the power conductors in the conduit. Ref.—Section 300-11(b) Exception 2

22. Underground water piping system must be at least 10 feet and be supplemented by an electrode type. Ref.—Section 250-50

23. Yes. Ref.—Section 210-70

24. Entire fixture. Ref.—Section 410-4

25. Yes, NO single conductors are OK; Type UF cable are not permitted. Ref.—Sections 225-6 and 339-3(b)(9)

PRACTICE EXAM 5

1. Yes, limited for corrosion and voltage isolation. Ref.—Section 318-3(e)

2. Ref.—Section 410-8(d)

    (a) 12-inch incandescent, 6-inch fluorescent

    (b) 12-inch incandescent, 6-inch fluorescent

    (c) 6-inch incandescent, 6-inch fluorescent

3. No, generally four or less No. 14 AWG fixture wires are not required to be counted. Ref.—Section 370-16(b)(1)

4. Yes, generally. Ref.—Section 110-26(c)

5. No. 4 copper. Ref.—Section 250-66 Exception 1

6. Yes. Ref.—Section 700-12(e), Exception

7. No, must be factory installed. Ref.—Section 550-23(b)(1)

8. Yes. Ref.—Section 680-25(d)

9. Yes, when smaller than $\frac{5}{8}$ inch must be listed. Ref.—Section 250-52(c)(2)

10. Both are acceptable. Ref.—Section 230-43

11. Exothermic welds. Ref.—Section 250-70

12. No. Ref.—Section 250-118 Exceptions 1 and 2

13. Yes. Ref.—Section 518-4(a)

14. No. Ref.—Section 430-21

15. Yes. Ref.—Section 351-23(a) and (b)

16. Yes, but bare conductors are permitted. Ref.—Section 230-41 and Exception

17. No. Ref.—Sections 300-3(c) and 725-54

18. No, only circuits over 250 volts to ground. Ref.—Section 250-97. 1999 *NEC®* change: See the new exception for listed boxes with a specified type of concentric knockouts that are acceptable.

19. Yes. 1999 *NEC®* New Exceptions would permit this practice in some instances without raceway protection. Ref.—Section 300-22(c)

20. Yes. Ref.—Section 300-13(b)

21. 60 amperes. Ref.—Section 430-72(b) Exception 1—Table 430-72(b) Column B

22. Yes. Ref.—Sections 410-56(f), 250-146

23. Code uses nominal voltages. Ref.—See Definition Article 100 or Section 220-2

24. No. 12 TW conductor in cable or raceway. Ref.—Table 310-16 = 25 amperes

    Four conductors in a raceway Table 310-52(b)(2)—four to six conductors in a raceway or cable shall be derated to 80% = 25 × 80% = 20 amperes the allowable ampacity.

    Ambient temperature correction factors shown at the bottom of Table 310-16 requires an additional deduction of .82 − 20 × .82 = 16.4 amperes. Section 210-21—outlet devices shall have an ampere rating not less than the load to be served. Table 210-21(b)(2) permits a 20-ampere branch-circuit for a maximum of 16 amperes. 16.4 exceeds this requirement. The conductor will carry the maximum allowable load. *Note:* See obelisk note at the bottom of Table 310-16 that limits the overcurrent protection on a No. 12 conductor to 20 amperes.

25. 36 × 36. Ref.—Section 328-10

PRACTICE EXAM 6

1. Yes. Ref.—Sections 250-97, 250-94

2. Six. Ref.—Sections 230-71(a), 230-40, Exception 1

3. (a) Yes. Ref.—Section 600-10(c)(2)

    (b) No. Ref.—Section 600-11—There is no exception

4. Yes. Ref.—Sections 545-2, 210-70 (a dwelling), 210-70(c) (other types of buildings)

5. Yes, if they meet all of the conditions. Ref.—Section 605-8

6. Yes. Ref.—Section 510-5(a), Article 500—Gasoline is a Group D, Class I hazard. Diesel is not. However, the sealing requirements are to prevent the passage of gases, vapors, and flames.

7. Yes. Ref.—Section 210-52(a)

8. (a) No. 2 copper or 1/0 aluminum. Ref.—Section 250-24 Base on 300 kcmil Service-Drop Table 250-66

(b) Bonded to the panelboard supplying each apartment size to Table 250-122 base and an overcurrent device 100 amperes No. 8 copper or No. 6 aluminum.   Ref.—Section 250-104

(c) No.   Ref.—Section 250-50—a supplementary grounding electrode only required where the underground water pipe is available. This installation is suggested by a nonmetallic water system.

9. Yes.   Ref.—Section 210-23(a)

10. No, generally. Yes, for interior portion of a one-family dwelling. Ref.—Section 680-25(c) Exception 3

11. No, where an equipment grounding conductor and not over four fixture wires smaller than No. 14 enter the box from the fixture.   Ref.—Section 370-16(a) Exception

12. Receptacle is not required specifically. Section 210-52(a) would require a receptacle near the wet bar. If located on counter top, a GFCI is required. Ref.—Sections 210-52(a), 210-8(a)(7). See 1999 *NEC*® change to Sections 210-52(a), 210-8(a)(7).

13. Read heading for Table 220-19. Note 3 permits add nameplates of the two ranges. Use Column C. Ref.—Table 220-19 and Note 3

14. 4.5 kW.   Ref.—Section 220-3(b)-(i). Section 220-18 is used for feeder load calculations.

15. 35 pounds.   Ref.—Section 422-18

16. Table 350-12 for ⅜ inch. For ½ inch through 4 inch use Table 1, Chapter 9.   Ref.—Section 350-12

17. No.   Ref.—Section 370-23(b), 370-23(a)–(g). *Note:* 300-11(a) the general method requirements give limited permission for branch-circuits. However, junction boxes are covered by Article 370.

18. No, explosion proof not required or acceptable. Must be approved for Class II locations.   Ref.—Section 502-1

19. No. THHN is not acceptable in wet locations. Ref.—Table 310-13 and Section 310-8

20. "Authority Having Jurisdiction."   Ref.—Section 90-4, also see Definition Article 100 "Approved"

21. Yes.   Ref.—Section 210-52(f)

22. Yes.   Ref.—Section 210-63

23. Column A is the basic rule to be applied when an exception does not apply.   Ref.—Section 430-72(b)

24. 83%.   Ref.—Duff & Herman, *Alternating Current Fundamentals, 4E*, Delmar, Thomson Learning, and Chapter 3 of this book.

25. Qualified yes "Metal Multioutlet Assembly." Ref.—Section 353-3

PRACTICE EXAM 7

1. Lighting circuit serving the area.   Ref.—Section 700-12(e)

2. Yes, if they are "Classified" for purpose and meet all conditions of Section 318-7(b).   Ref.—Section 318-7(a) and (b)

3. No.   Ref.—Section 352-22

4. If the required work space is unobstructed or doubles.   Ref.—Sections 110–110-16(c), Exceptions 1 and 2

5. No, the anti-short bushing is not required. The conductors in MC cable are wrapped in a nylon wrapping. AC cable conductors are individually wrapped in kraft paper.   Ref.—Sections 333-9 Armored Cable, 334-12 Type MC Cable

6. Generally, no, if exposed to physical damage. Ref.—Section 331-4(3) and (7). *Note:* If these conditions do not exist, then ENT could be installed.

7. 18 inches.   Ref.—Section 300-5

8. Large capacity multibuilding industrial installations under single management.   Ref.—Section 225-32 Exception 1

9. No for Class 2 Div. 1, Yes for Class 2 Div. 2. Ref.—Section 502-4(a) and (b)

10. No.   Ref.—Section 410-31

11. Grounded (neutral) conductor must be insulated generally.   Ref.—Sections 310-12; 250-140 and 142. The rules for an insulated neutral or grounded conductor for a feeder, although not clearly written, the service neutral is often bare and Section 250-140 permits dryers and ranges originating in the service panel to be Type SE, which has a bare grounded conductor.

12. No. Classifying structures are the responsibility of the building inspector or the fire marshal generally. They rely on the adopted building code or the NFPA 101 Life Safety Code.   Ref.—Article 518

13. No.   Ref.—Section 220-22, Table 310-15(b)(4)

14. None.   Ref.—Table 310-16 Ambient Temperature Correction factors at bottom of table.

15. Table 370-16(b)

    Example: An outlet box contains a 120-volt, 20 amp single receptacle and a 120-volt 30 amp receptacle. The 20-ampere receptacle is fed with a 12-2 W/GRO NM cable that extends from the outlet on to additional outlets located elsewhere. The 30-ampere receptacle is fed by an individual branch-circuit with 10-2 W/GRO NM cable. What size box is needed?

    Answer: 370-16(a)

12-2 W/GRO Feed	2-12 @ 2.25 cu in	4.5 cu in
12-2 extending on	2-12 @ 2.25 cu in	4.5 cu in
10-2 W/GRO feed	2-10 @ 2.5 cu in	5 cu in
Receptacle	2-10	5 cu in
GRO	-10	2.5
Clamp	-10	2.5
		23.5

    A box with 23.5 cubic inches is required.

16. 10 feet.   Ref.—Section 230-24(b)

17. No, it is considered by the *NEC*® as "Other Space Used for Environmental Air."   Ref.—Section 300-22(c) and Exceptions, Definition of "Plenum" Article 100

18. No, this area must be unobstructed.   Ref.—Section 410-8(a)

19. Equipment grounding conductor required.   Ref.—Sections 250-134, 250-136

20. Yes.   Ref.—Section 410-4(a)

21. By interpolation.   Ref.—Section 430-6(a)

22. 53%.   Ref.—Chapter 9, Table 1 and Note 6

23. Yes.   Ref.—Table 310-15(b)(6)

24. No, motors are computed in accordance with Article 430.   Ref.—Section 220-14, Section 680-3

25. Yes, if supplied with electricity.   Ref.—Sections 210-8(a)(2), 210-52(g)

PRACTICE EXAM 8

1. Yes, generally.   Ref.—Section 410-73(e)

2. As per Section 547-8 in compliance with Article 250.   Ref.—Section 250-32(e) (FPN) and Section 547-8

3. Yes, if known, and they occupy a dedicated space with the individual branch-circuit.   Ref.—Table 220-19 "Heading." If they are plugging into one of the "small appliance" outlets, then they would not have to be added to the load calculations.

4. Yes, for controlled starting.   Ref.—Section 430-82(b)

5. No.   Ref.—Sections 553-2 and 553-4

6. No.   Ref.—Article 600, Section 210-6(a)

7. Yes.   Ref.—Section 210-21

8. Yes, limited application.   Ref.—Articles 517-12, 13, 30

9. No, generally.   Ref.—Section 240-3 and Section 310-15

10. Yes.   Ref.—Sections 240-83(d), 380-11

11. There shall be two deductions for the largest conductor connected to each device, and one deduction of the largest conductor entering the box for each of the following: the clamps, the hickeys, all of equipment grounding conductors, etc.   Ref.—Section 370-16(b)(1)

12. No.   Ref.—Section 250-94

13. No, the cable can be fished; however, Article 370 does not permit the fished NM cable without a box clamp.   Ref.—Section 336-18 Exception 1, Section 370-17(c) Exception

14. 150 amp will meet all conditions required for a general use switch. Ref.—Section 440-12, Section 430-109(d)(3)

15. No.   Ref.—Section 350-10(2) and (5)

16. I P/E   I = 10,000/240   I = 41.667

    R = E/I   R = 240/41.667   R = 5.76

    P = IE   P = 208 × 36.1   P = 7.511 kW

    I = E/R   I = 208/5.76   I = 36.1 amperes

    36.1 amperes—Chapter 3, Ohm's Law

$$I = P/E \quad I = 10,000/240 \quad I = 41.667$$

$$R = E/I \quad R = 240/41.667 \quad R = 5.76$$

$$I = E/R \quad I = 208/5.76 \quad I = 36.1 \text{ amperes}$$

17. Yes. Ref.—Section 680-20(a)(1), Section 680-5(a), (b), and (c)

18. Chapter 9, Table 1 unless durably and legibly marked by manufacturer in cubic inches. Ref.—Section 370-16(c)

19. By the markings on the fixture by the listing agency. Ref.—Section 410-65(c), Exceptions 1 and 2

20. Conductors only. Ref.—Section 240-21

21. Table 110-16a, Condition 3. Ref.—Table 110-16a

22. Yes, if cord and plug connected but not required or advisable. Generally, these appliances are wired to the requirements in Article 422. Ref.—Section 210-52(b) Exceptions 1–5 and (2)

23. 10 feet. Ref.—Section 680-8(1)

24. Yes, no depth requirement but is required to be protected from severe physical damage. Ref.—Sections 250-64 and 250-120

25. Water meter and all insulating joints. Ref.—Section 250-81(a)

PRACTICE EXAM 9

1. Yes, but not advisable. One should isolate any possibility of imposing faults on those circuits. Ref.—Section 680-5(b), Section 680-20(a)

2. (a) Yes. Ref.—Section 250-21, Section 250-24(a) and (b)

   (b) In each of the paralleled raceways. Ref.—Section 300-3(b), Section 250-24(a) and (b)

   (c) No. 2 copper in each conduit. Ref.—Section 250-24, Section 250-66

3. 21 inches × 24 inches × 6 inches. Ref.—Section 370-28(a)(2)

4. Yes, by Exception. Ref.—Section 373-8

5. Yes. Ref.—Section 348-13 Exception

6. Either system installation is okay where a three-pole transfer switch is used. Grounding electrode is not required for this installation. Ref.—Section 250-23(d), Section 250-30

7. No. Ref.—Section 300-5(i), Exception 2, although grouping of the conductors is preferred.

8. This installation is a violation; you cannot mix these conductors, unless the control is a Class 1 or power circuit. Ref.—Section 300-3(c)(1)

9. Permit, however, it should not be located directly over the switchboard for the best coverage. Ref.—Sections 384-4(FPN-2), 110-26(r)(1)(2)

10. Yes. Ref.—Section 250-50. Steel can be bonded to the water pipe within 5 feet of where the water pipe enters the building.

11. 20-ampere breaker. Ref.—Section 310-16 obelisk, Section 210-22

12. Yes, however, the receptacles still must be installed at direct grade access and be GFCI protected. Ref.—Section 210-8a(3), 210-52(e); 1999 *NEC®* change.

13. Yes, if listed for 2. Ref.—Section 110-3b, Section 373-5(c)

14. No. Ref.—Section 230-42(a), (b), and (c)

15. Yes. Ref.—Section 430-109(a)(1)

16. Article 310 Ref.—Table 310-15(b)(6)

17. No, neutral is not used at the switch. Ref.—Section 200-7(c)(2), Section 300-3(b)

18. Yes, but where they enter the building they must be enclosed in a raceway. Ref.—Section 339-3(a)(2)

19. Yes, area may be declassified by AHJ where there are four air changes per hour. Ref.—Sections 511-7(a), 511-3(a)

20. No. Neutral must be capable of carrying the maximum unbalanced load. Ref.—Section 250-140, Section 210-22

21. Yes. Ref.—Section 350-10

22. Maybe AHJ is the building inspector. Ref.—Local adopted building code—Section 518-2(c)(FPN)

23. Yes, but it must be connected to the same circuit as supplies the lighting for the area to be served. Ref.—Section 700-12(e)(4)

24. No. This wet location is defined in Section 517-3.

25. Yes; No.   Ref.—Section 422-12

PRACTICE EXAM 10

1. Yes, but storage of combustibles should be considered.   Ref.—Part B Article 410

2. Hot tubs and spas are defined in Article 680. Ref.—Section 680-4

3. No.   Ref.—Section 511-4, Section 501-5(a)(4) Exception

4. 150, No. $P = EI = 480 \times 104 = 49,920$ single phase

   $49,920/480 \times 173 = 60.12$ amperes

   $60.12 \times 250\% = 150$

   Ref.—Section 450-3(b)(2)

5. Each section 5 feet or fraction thereof shall be calculated at 180 VA in other than dwellings. Ref.—Section 220-3(b)(8)

6. Yes.   Ref.—Part I Article 430

7. Yes, possibly.   Ref.—Section 430-72 and Exceptions

8. Yes, except in an industrial facility.   Ref.—Section 364-11

9. No.   Ref.—Section 339-3b(1), (9), Section 300-5(d)

10. Yes.   Ref.—Sections 250-20 and 24(a) and (b)

11. No.   Ref.—Section 250-30 covers the requirements for a separately derived system. Nearest building still is preferred.

12. No, you cannot tap a tap. A tap is defined in Section 240-3(e).   Ref.—Section 240-3(e)

13. Yes.   Ref.—Part C, Article 424, Section 424-20(b)

14. No. 2.   Ref.—Chapter 9, Table 8, Table 250-66

15. Yes. However, consideration must be made for corrosive elements. Ref.—Sections 300-6, 501-4a, 346—Non-Ferrous Rigid Metal Conduit.

16. No.   Ref.—Section 700-12(f)

17. No.   Ref.—Section 250-140. See 1999 *NEC®* change for new installations.

18. Yes, but there are installation requirements for disconnecting overcurrent.   Ref.—Sections 230-82, 695-3

19. Bond services together.   Ref.—Section 250-58

20. No.   Ref.—Sections 680-71, 410-4(a)–(d)

21. No.   Ref.—*NEC®*. Most requirements for GFCI protection appear in Section 210-8. Other requirements appear in Articles 305, 511, 517, 550, 551, 553, 555, and 680.

22. No.   Ref.—Section 410-28(d), Section 410-31

23. Yes.   Ref.—Section 331-3(5). The 15-minute finish rating is a wall finish rating established by Underwriters Laboratories and can be found in the "UL Fire Resistant Directory."

24. No.   Ref.—Article 370, Section 370-16(a) and (b), Table 370-6(b). $4 \times 2\frac{1}{8} = 30.3$ cubic inches. No. 8 conductor = 3 cubic inches. Therefore, ten No. 8 conductors would be allowed without considering devices or other appertances.

25. They must be fastened.   Ref.—Section 410-16(c)

PRACTICE EXAM 11

1. Yes.   Ref.—Section 339-3(a)(4)

2. No.   Ref.—Section 680-20(a)(1)

3. Yes.   Ref.—Section 700-12(e) Exception

4. No.   Ref.—Section 300-22(c), Exception. Perpendicular, yes; parallel, no.

5. No, they must be equipped with a factory-installed GFCI.   Ref.—Section 600-11

6. In each disconnect means.   Ref.—Section 250-64(c)

7. No.   Ref.—Section 250-64(c) Exception 2

8. As a wireway in accordance with Article 362. Ref.—Sections 362-5, 362-7

9. Yes.   Ref.—Section 250-148.

10. No. It can be insulated or covered. Ref.—Section 680-25(c)

11. Yes. This installation is not in violation of the Code. Ref.—Sections 240-3, 310-16, 220-10, 384-13.

    Section 240-3(b)—The 400-ampere breaker properly protects 500 kcmil conductors that permit the next higher overcurrent device.

    Section 220-10—Conductors 500 kcmil are large enough to carry the 180-ampere load.

    Section 384-13—It is a power panel, not a lighting and appliance branch-circuit panelboard.

12. Yes, unless not required by NFPA 99. Ref.—Section 517-25 (FPN)

13. Generally, no—Table 220-11, and 220-13. Ref.—Section 220-3(a), (b), and (c), Section 220-10(a), (b)

14. Yes, No. Ref.—Section 517-3 definition of patient care areas in hospitals. Ref.—Section 517-13(a)

15. Yes, there are no conditions. Ref.—Section 600-7, Section 250-112

16. No. Ref.—Section 250-30

17. Yes. Ref.—Section 310-4

18. They must be derated to 80%. Ref.—Table 310-15(b)(2)

19. Yes. Ref.—Section 210-6(c)

20. Yes. Ref.—Definition "Accessible Readily," Definition of "Direct Grade Access"—Section 210-52(e). See 1999 *NEC*® change.

21. Yes. Section 210-52(d), Section 210-11 Exception. New change 1999 *NEC*®.

22. Yes, if they do not exceed allowable load of circuit. They are not continuous loads. Ref.—Sections 210-19(c), 210-22(a), 210-23

23. No. Ref.—Section 210-52(g) only applies to dwellings (see Definition Article 100)

24. Yes, Yes. Ref.—Section 210-70(a) Exception 2

25. Overcurrent device, Yes. Ref.—Section 220-10(a) and (b)

PRACTICE EXAM 12

1. Only branch-circuits. Ref.—Sections 220-15 and 424-(3)(d)

2. No. Table 220-19 Heading Column A applies to all notes as applicable, Columns B and C apply to Note 3. Ref.—Same

3. Yes. Ref.—Sections 225-32, 225-34, 230-71 and 72

4. Location is not specified; where bonded to different electrodes they must then be bonded together. Ref.—Section 250-58

5. Yes, but may be wired as nonhazardous. Ref.—Section 511-3(d) and Section 514-1.

6. ½ inch. Ref.—Section 410-66(a)

7. System ground. Ref.—Section 810-21(b)

8. No, Section 90-3. Use Chapters 1–4. Ref.—Section 511-3(d)

9. Yes. Ref.—Section 250-97 Exception. See new exception in 1999 *NEC*®.

10. Yes. Ref.—Section 680-25(c)

11. No, only as taps (from raceway system to fixture). Ref.—Section 350-10(2)(c)

12. No. Ref.—Section 680-20(c)

13. Yes. Ref.—Section 680-40 Exception 2 and 680-41(c)

14. Section 210-70(a) specifically states *at outdoor entrances or exits*. Ref.—Section 210-70(a)

15. 65%—Column C, Note 3—add 3 + 6 = 9 × 65% = 5.85 kW

16. Yes. Ref.—Section 210-52(f). At least one but only for the laundry area

17. Yes, all. Ref.—Section 210-8a(1)

18. No, except when using Class 1 or power circuits for control. Ref.—Section 300-3(c)(1), Section 725-54(a)(1)

19. Yes. Ref.—Section 800-33

20. Bonding jumper shall be the same size as the grounding-electrode conductor. Ref.—Sections 250-28 and 250-102

21. Yes, for wet bar sink; no, for laundry sink. Ref.—Section 210-8(a)(7)

22. 25 ft. tap = No. 1 THW 20 ft. tap = No. 10 THW. Ref.—Section 240-21, Table 310-16 and obelisk note at bottom of table

23. No.   Ref.—Section 210-70a. At least one switch

24. 75°C, because there are no circuit breakers rated 90°C. Ref.—Section 110-14c, Section 310-15c

25. Those areas determined by governing body of facility.   Ref.—Section 517-3 Definitions, Patient Care Areas of a Hospital

PRACTICE EXAMINATION

1. C. Ref.—Article 100, Definition
2. A. Ref.—Section 310-11(c)
3. E. Ref.—Section 725-23
4. D. Ref.—Section 373-8
5. B. Ref.—Section 240-51(a) and 240-53(a)
6. D. Ref.—Section 760-51(a)
7. C. Ref.—Chapter 9, Note 4 to Table 1
8. D. Ref.—Section 240-80
9. C. Ref.—Section 720-5
10. D. Ref.—Table 402-5
11. A. Ref.—Chapter 9, Table 8, Col 6
12. B. Ref.—Section 760-54(c)(1)
13. B. Ref.—Section 110-1
14. D. Ref.—Section 346-12(b)(2) and Table 346-12
15. D. Ref.—Article 100, See Definition Switch
16. D. Ref.—Section 725-8(b)
17. B. Ref.—Section 250-12
18. C. Ref.—Article 100, Definition
19. D. Ref.—Section 110-27
20. B. Ref.—Section 110-2
21. B. Ref.—Section 90-3
22. B. Ref.—Article 100, Definition
23. B. Ref.—Section 725-54(a)(1)
24. B. Ref.—Article 100, Definition
25. A. Ref.—Section 200-6
26. B. Ref.—Section 110-19
27. A. Ref.—Section 424-39
28. C. Ref.—Section 660-5
29. A. Ref.—Section 225-14(c)
30. C. Ref.—Section 110-26(e)
31. A. Ref.—Section 220-3(b)(8)(2)
32. A. Ref.—Section 810-11 Exception
33. B. Ref.—Section 230-79(c)
34. A. Ref.—Section 430-89
35. D. Ref.—Section 225-4
36. D. Ref.—Section 110-53
37. D. Ref.—Section 230-84(b) Exception
38. A. Ref.—Section 210-23(a)
39. D. Ref.—Section 374-6, copper (assumed) is 1,000 A per square inch. $4 \times \frac{1}{2} = 2$ square inches, $1,000 \times 2 = 2,000$ amps max.
40. D. Ref.—Section 305-7
41. D. Ref.—Section 430-32(b)(1)
42. D. Ref.—Section 550-11(a)
43. C. Ref.—Section 600-9
44. D. Ref.—Section 820-40(a)(1, 3, 5)
45. E. Ref.—Section 110-32
46. C. Ref.—Section 517-15(a)
47. D. Ref.—Section 320-8
48. A. Ref.—Article 100, Definition
49. A. Ref.—Section 225-6(a)(1)
50. C. Ref.—Section 600-32(b)

FINAL EXAMINATION

1. C. Ref.—Section 800-12b
2. B. Ref.—General mnowledge
3. D. Ref.—General knowledge, *IEEE Dictionary*
4. A. Ref.—Chapter 3
5. A. Ref.—Chapter 3
6. C. Ref.—Chapter 3

7. C. Ref.—Chapter 3

$$30P = I (E \times 1.73) \quad P = 83 \text{ kW}$$

8. A. Ref.—Chapter 3

9. A. Ref.—Chapter 3

10. C. Ref.—Chapter 3

$$16 \times 4 = 64 \times 40 = 10 \times 4 = 40$$

11. A. Ref.—Section 250-102(c)

Note: 12½ minimum or not smaller than required in Table 250-94, individual bonding jumpers could be run from each conduit sized 1/0

12. A. Ref.—Chapter 9, Tables 1, 5, and 4

2/0 = .2265 sq inches × 2 = .4530

No. 1 = .1590

.4530 + .1590 = .6120 Table 4 = 1½ inches

13. B. Ref.—Table 250-66

14. C. Ref.—230-95(a)

15. B with Exception. Ref.—Section 230-9

16. B. Ref.—Duff & Herman, *Alternating Current Fundamentals, 4E*, Delmar, Thomson Learning

17. B. Ref.—Table 310-15(b)(2)

18. A. Ref.—Note 8, Ampacity Tables 0–2000 Volts

19. B. Ref.—Chapter 9, Tables 1, 5, and 4

Solution: Table 1 over 2 cond. limits the raceway to a 40% fill.

Table 5: #4 THWN = .0824 sq. in.
⅒ THWN = .2223 sq. in.

$$3 \times .0824 = .2472$$
$$4 \times .2223 = \underline{.8892}$$
$$1.1384 \text{ sq. in.}$$

Table 4 = 2 in. IMC

20. B. Ref.—Appendix C, Table C-10

21. C. Ref.—Section 230-70(a)

22. C. Ref.—Appendix 1

23. D. Ref.—Appendix 1

24. B. Ref.—Section 430-52

25. D. Ref.—Article 430, Sections 430-37, 38, and 39, Table 430-37.

26. B. Ref.—Section 430-126

27. A. Ref.—Article 430 Part J

28. D. Ref.—Section 430-89

29. D. Ref.—Section 250-140 in existing installations only. See 1999 *NEC®* changes.

30. B. Ref.—Section 250-110(5)

31. C. Ref.—Section 250-52(c)(3)

32. D. Ref.—Article 430

33. D. Ref.—Section 225-18

34. C. Ref.—Section 230-95(c)

35. A. Ref.—Section 430-24

Largest $5.6 \times 1.25 = 7$
$4.5 \times 1.00 = 4.5$
$4.5 \times 1.00 = \underline{4.5}$
16 amperes—minimum ampacity of conductors

36. B. Ref.—Section 380-8(a)

37. D.

38. D.

39. A.

40. B.

41. C.

42. A.

43. D.

44. B.

45. A.

46. B.

47. D.

48. A.

49. B.

50. C. Ref.—Ohm's Law, 100% load

1999 CODE CHANGE QUIZ ONE

1. No. Ref.—Article 100 definition of service, it must be supplied from the serving utility.

2. Article 110 Part D. Ref.—Table of Contents

3. 6 feet, lower. Ref.—New Section 110-26(f)

4. Section 110-54(a) and 110-54(b). Ref.—Same

5. Three continuous white stripes. Ref.—Section 200-6

6. Yes. Ref.—210-8(a)(2)

7. *An arc-fault circuit interrupter is a device to provide protection from the effects of arc faults by recognizing characteristics unique to arcing and by functioning to de-energize the circuit when an arc-fault is detected.* Ref.—210-12(a)

8. No, that would be a feeder by definition. Ref.—Article 100 definitions

9. False. Ref.—Section 110-26(a)(3). Equipment located above or below other electrical equipment is permitted to extend up to 6 inches beyond the front of other electrical equipment.

10. They were moved to Article 110. Ref.—110-26(f)(1)(a)

11. They were moved to Article 110 Part C. Ref.—Same

12. Three continuous white stripes on other than green insulation along its entire length. Ref.—Section 200-6

13. GFCI (ground-fault circuit interrupter protection for personnel). Ref.—Section 210-8(a)(2)

14. January 1, 2002. Ref.—Section 210-12(b)

15. No. Ref.—Section 210-52(b)(3). Some dwellings have more than one kitchen and this new requirement was added to reduce overloading.

16. Adjacent, 36. Ref.—Section 210-52(d). The required receptacle(s) must be located on the wall adjacent to the basin to prevent draping a cord across the passage.

17. Two, readily accessible. Ref.—Section 210-60(b)

18. 150 volt-amperes. Ref.—Section 220-12(b)

19. 12 feet. Ref.—Section 225-18

20. 100 amperes. Ref.—Section 225-39(c)

1999 CODE CHANGE QUIZ TWO

1. Section 230-40 Exception 4. Ref.—Same section

2. 230-79(c). Ref.—Same section

3. No, Section 240-3(d). Ref.—Section 240-3(d)

4. Section 240-3(e). Ref.—Section 240-3(e)

5. Unlimited length. Ref.—Section 240-21(b)(5)

6. New Part H to Article 240, Section 240-91. Ref.—Same

7. They are required to be GFCI protected. Ref.—Section 525-18(a)

8. First paragraph of Section 240-21, which states: *No conductor supplied under the provisions of Sec. 240-21 (a) through (g) shall supply another conductor except through an overcurrent device.* Ref.—Section 240-21, second sentence, first paragraph

9. No, the exception to 220-16 allows this load to be excluded from the calculation. Ref.—Sections 210-52(b)(1) Exception 2 and 220-16 Exception

10. Yes, they would have to be accessible; however, only two would have to be readily accessible. Ref.—Section 210-60(a) and (b)

11. No, they would still be accessible as defined by the Code. Ref.—Article 100 definition of accessible and readily accessible

12. Where the generator is a separately derived system. Ref.—Section 250-20(d) and (FPNs) 1 and 2

13. There is no set distance other than it must be within 36 inches of the outside edge of each basin and it must be located on the wall immediately adjacent. Ref.—Section 210-52(d)

14. Yes, an exception has been added to state that if the circuit supplies only a single bathroom then it is permitted to supply other equipment within that single bathroom. Ref.—Section 210-11(c)(3) and Exception

15. Yes, in places of assembly Article 518. Ref.—Section 518-3 Exception

16. Listed. Ref.—Section 695-10

17. No. Ref.—Section 547-9(a)

18. No, an equipment grounding conductor in accordance with Section 250-118 is required, and the grounded (neutral) cannot be bonded to the grounding-electrode system where there are common metallic paths between the two buildings. Ref.—Section 250-32(b)(1) and (2)

19. Yes, if it meets the conditions of Section 373-5(c) Exception.   Ref.—Same

20. Yes, if they have a floor located at or below grade and are not intended as habitable, and are limited to storage areas, work areas, or similar use. Ref.—Section 210-8(a)(2)

1999 CODE CHANGE QUIZ THREE

1. Yes, or ground-fault protection is installed in accordance with Section 250-122(f)(2).   Ref.—Same

2. A power panelboard is one having 10% or fewer of its overcurrent devices protecting lighting and appliance branch-circuits.   Ref.—Section 384-14(b)

3. Yes, in accordance with Sections 110-14, 300-5(e), 300-13, and 300-15.   Ref.—Section 230-46

4. Yes, for feeders or branch-circuits.   Ref.—Section 305-4(b) and (c)

5. Yes, in size ½ through 1 inch as a listed manufactured prewired assembly.   Ref.—Section 331-3(8)

6. Yes, only one is permitted generally. Additional feeders or branch-circuits are permitted in accordance with Section 225-32(a) through (e).   Ref.—Same

7. Type AC cable with an insulated equipment grounding conductor sized to Table 250-122 is now allowed.   Ref.—Section 518-4(a)

8. Yes, they are not permitted to be installed in a face-up position in the counter tops or work spaces.   Ref.—Sections 210-8(a)(7) and 210-52(c)(5)

9. The industrial portion of a facility meeting all the conditions of Section 240-91 (1) through (3). Ref.—Same

10. Listed tamper resistant or employ listed tamper-resistant covers.   Ref.—Section 517-18(c)

11. Yes, if the light fixture does not exceed 6 pounds. Ref.—Section 370-27(a) Exception

12. At least 3 inches of conductor outside the opening. Ref.—Section 300-14

13. No, the building codes may also prohibit combustible wiring methods.   Ref.—Section 336-5(a)(1)

14. It is not required for all buildings; industrial buildings are exempted. However, where the busway vertical riser penetrates two or more floors, a 4-inch minimum curb must be installed to retain liquids.   Ref.—Section 364-6(b)(2)

15. Two minimum where a made electrode is used and they have not been tested for compliance with Section 250-56.   Ref.—Section 250-50(a)(2)

16. No, only medium- and low-power systems are covered by Article 830.   Ref.—Section 830-4 and Table 830-4

17. Where a general care patient bed location is served from two transfer switches on the emergency system they are not required to have circuits from the normal system.   Ref.—Section 517-18(a) Exception 3

18. Yes, they could all be main power feeders if feeding lighting and branch-circuit panelboards. Ref.—Section 310-15(b)(6)

19. Yes.   Ref.—Section 210-8(a)(3)

20. New Part D of Article 370.   Ref.—Same

Index